Oil & Gas Company Analysis Petroleum Refining & Marketing

Alfonso Colombano
Alberto Colombano

ISBN-13: 978-1546850199

ISBN-10: 1546850198

Printed by CreateSpace, An Amazon.com Company

www.oilgascompanyanalysis.com

Disclaimers

All data discussed and used in this book has been referenced and sourced from **publicly** available materials, such as companies' forms 10-K and 20-F, annual reports, internet websites, news articles, company reports and news releases. Reader should verify accuracy by checking cited references.

The recommendations, advice, analysis, descriptions, methods and calculations presented in this book are for educational and illustration purposes only. The authors shall not be liable for any investment loss or any damage that results from the use of any of the material in this book.

This book is for informational purposes only and does not constitute an offer to sell, a solicitation to buy, or a recommendation for any security, nor does it constitute an offer to provide investment advisory or other services by the authors of this book. No reference to any specific security constitutes a recommendation to buy, sell or hold that security or any other security. Nothing in this book shall be considered a solicitation or offer to buy or sell any security, future, option or other financial instrument or to offer or provide any investment advice or service to any person in any jurisdiction. Nothing contained in this book constitutes investment advice or offers any opinion with respect to the suitability of any security, and the views expressed in this book should not be taken as advice to buy, sell or hold any security.

While every effort has been taken in the preparation of this book, the authors assume no responsibility for errors or omissions. Due to the large volume of public information considered (eight companies' published reports in various venues) metrics will differ. We apologize in advance for any errors that may be found in this book. This book is for educational purposes only. If you find an error or omission, please let us know at book@oilgascompanyanalysis.com.

Alfonso Colombano, Alberto Colombano or the book's publishers shall not be liable for any damages whatsoever, and in particular the authors shall not be liable for any special, indirect, consequential, or incidental damages, or damages for lost profits, loss of revenue, or loss of use arising out of or related to this book or the information within, whether such damages arise in contract, negligence, tort, under statute, in equity, at law, or otherwise, even if the authors have been advised of the possibility of such damages.

Table of Contents

Contents 5

About Alfonso Colombano 11

About Alberto Colombano 13

Acknowledgements – Alfonso Colombano.......................... 15

Preface 17

Chapter I – Introduction 19

    Audience 20

    What readers will learn from this book 20

    Why this book? 21

    What is oil? 21

    What is natural gas?.. 27

    What are NGLs?.. 28

    What is Upstream, Midstream and Downstream? 29

    Downstream Overview 31

    Other businesses in Downstream............................... 32

    Oil & Gas Companies in this book............................ 34

    Units and Conversion... 37

Chapter II - Financial Analysis Overview.......................... 41

    Financial Analysis 42

    Financial Statements 42

    Company Structure 47

    Financial Metrics Overview 50

    Market Capitalization... 51

    Enterprise Value... 51

    Total Consolidated Revenues.................................... 52

Earnings 53

Adjusted Earnings 54

Gross Margin & Gross Operating Margin 55

Gross Margin Percentage 55

Net Margin...56

EBITDA 56

EV/EBITDA or EBITDA Multiple..57

Return on Capital Employed (ROCE)..58

Cash Return on Capital Employed (CROCE) 59

Return on Average Equity (ROE) 60

Dividends or Distributions per Share/Unit.................................61

Dividend Yield..62

Capital Expenditures..63

Cash Flow from Operations (CFO) 64

Free Cash Flow (FCF) 65

Percent of Cash Flow from Operations devoted to Dividends or Distributions...65

Total Shareholder Return (TSR) 66

Debt to Equity Ratio..67

Current Ratio..68

Working Capital 69

Price Earnings Ratio...70

Interest Coverage Ratio 70

Chapter III – Petroleum Refining 71

Refining...72

Definitions 74

Why do we need Refining? 82

Refining Process 86

Types of Refineries..95

Roles in a Refining Organization..96

Global Refining Capacity & Throughput.................................99

Refining Key Metrics Overview..104

Total Crude Oil Processed..104

Total Throughput Volumes...104

Refined Product Yields..105

Total Refining Capacity..105

Refinery Utilization 106

Clean Product Yield..107

Refinery Complexity Index...108

Earnings per barrel..109

Cash per barrel...110

Market Crack Spread...111

Realized Crack spread..112

Cash Flow Profile of a Refining Asset...................................114

Business Cycle in Refining..115

Why invest in Refining?..117

Chapter IV – Petroleum Marketing & Trading...........................119

Marketing & Retail Overview...120

Businesses within Marketing & Trading.................................123

Wholesale Marketing...123

Retail Marketing..123

Branded...125

Unbranded..127

Comparison – Unbranded vs. Branded 128

Marketing & Trading Supply Chain 129

Fuel Taxes...133

Roles in a Marketing & Retail Organization 134

Supply & Trading...135

Roles in a Supply & Trading organization 147

Marketing, Retail & Trading Key Metrics Overview 149

Total Refined Product Sales 150

Total Number of Stations 150

Fuel Volume Sales 151

Fuel Volume per Station 151

Gross Margin per Gallon or Liter 151

Merchandise Sales as Percent of Total Sales 152

Earnings per Barrel 152

Cash per Barrel 153

Business Cycle in Marketing & Trading 153

Why Invest in Marketing? 154

Chapter V – North America 155

ExxonMobil Downstream (XOM) 157

Tesoro Corp (TSO) 167

Delek US Refining 175

Chapter VI – Europe, Russia and CIS 185

Rosneft (OJSCY) 187

Total S.A. (TOT) 197

Chapter VII – Asia & the Middle East 205

Saudi Aramco (SA) 207

Petrochina (PTR) 213

Chapter VIII – Latin America 219

Petrobras (PBR, PBR-A) 221

Comparative Tables 231

CT1 – Refining Capacity and Throughput volumes 231

CT2 – Assessed Available Capacity as of January 2016 232

CT3 – Adjusted Earnings, Cash per Barrel and Clean Product Yield ... 233

CT4 – Alphabetical List of Companies 234

CT5 - Number of Vehicles per Country 235

CT6 – Vehicles per 1,000 People... 237

Index 239

Glossary 242

About Alfonso Colombano

Alfonso is the author of the bestseller *Oil & Gas Company Analysis: Upstream, Midstream & Downstream,* published in 2015, which has been used as a teaching tool in the oil & gas industry worldwide. Alfonso has worked for two International Oil Companies (IOC's) and one refining company in analytical, commercial & financial roles. Alfonso is a graduate of the University of Houston where he was active in the Energy Association. Alfonso's expertise encompasses all sectors of the oil & gas industry including upstream, midstream and downstream, which led him to create, write and publish three books on the topic. Alfonso is an avid reader of Forms 10-K and 20-F, financial statements and other external company publications.

Alfonso's expertise encompasses the financial & operational analysis of oil & gas companies, having developed several analytical models for M&A activity, commercial analysis as well as the model used for this book. Alfonso is a well-seasoned public speaker and has been featured at the University of Houston and the *Global Energy Leaders Podcast.* In addition, Alfonso is a subject matter expert in SAP® ERP and the U.S. natural gas industry. Alfonso enjoys reading books about the oil & gas industry and playing golf in his free time.

Alfonso can be reached by email at alfonso@oilgascompanyanalysis.com or by following his LinkedIn page at www.linkedin.com/in/alfonsocolombano

About Alberto Colombano

Alberto Colombano is an Information Technology professional, specializing in SAP® ERP. He has studied the mining and oil industries since 2005 along with economic analysis and worldwide monetary policy.

He graduated from University of Houston-Downtown with a Bachelor's degree in Computer Information Systems (CIS). He is multilingual (English, Italian, Spanish and German). He has extensive experience in SAP® ERP modules, such as FI/CO, MM, SD, PS and other modules.

He and his brother Alfonso read and study companies' financial reports such as 10-K and 20-F forms and other publicly available reports for the investment public. He helped with data processing and conversion by creating several custom applications & utilities.

Alberto can be contacted at alberto@oilgascompanyanalysis.com

Acknowledgements – Alfonso Colombano

"Learn everything you can, anytime you can, from anyone you can - there will always come a time when you will be grateful you did." Sarah Caldwell

Without the immeasurable aid, encouragement and support of my wife Indira, my parents and my siblings Alberto and Alfredo who encouraged and assisted me, I would never have completed this lengthy effort.

Additionally, I express my sincerest appreciation to Dr. Bill Leffler and his excellent collection of *Nontechnical Books*. Dr. Leffler has been a source of inspiration for me to write books ever since I started in the Oil & Gas industry several years ago. I want to also express my gratitude to my good friends at the University of Houston, Bauer College and the Energy Association for having broadened my vision about the industry and having invited high caliber presenters from Oil, Gas and Energy related firms throughout my studies at the University of Houston.

Last, but not least, my sincerest thanks to all the great colleagues, supervisors, managers, operations personnel, and mentors that I have had the good fortune to have worked with in the energy industry throughout my career. These great colleagues and mentors encouraged me from the beginning to learn more about our industry, the interconnectedness of the different sectors, the importance of focusing on fundamentals and the potential of the industry to grow long-term. I have been quite blessed to have been surrounded throughout my career, including at university and at work, with brilliant people who have influenced my views and have helped me tremendously.

Without the help of all of you, I could not have completed this monumental task. Thanks everybody!

Alfonso Colombano
Houston, Texas, USA

July 2017

Preface

"What you get by achieving your goals is not as important as what you become by achieving your goals." – Zig Ziglar

Dear reader,

Thank you for choosing this book. This is the third book I have published about the oil & gas industry. In this particular book, I'm looking forward to share more with readers about the petroleum refining & marketing business. From providing the fuel that we use in our cars, to transporting goods across the world, to producing the plastics that we use in our mobile phones, we can see *every day* how vital the petroleum refining & marketing business is to the world's economy.

The goal of this book is provide an *in-depth* analysis of the Refining & Marketing sector, profile various companies across refining & marketing as well as take a deeper dive into the metrics, analysis, and processes that make up these two businesses.

This book *builds on* the fundamental concepts from the first Oil & Gas Company Analysis book and examines them at *a much higher* level of detail. There were several changes implemented in this book:

- Improves and adds additional financial metrics, such as Enterprise Value, EBITDA, working capital, gross margin and other applicable financial metrics.
- Takes a deeper dive into petroleum refining processes, what impacts a refinery's economics, as well as refining & marketing driven metrics.
- Provide more insight into the petroleum marketing and trading activities associated with crude and refined products
- New global companies with varied business models in this industry are showcased.

Petroleum Refining & Marketing is a multifaceted, vibrant and dynamic industry. As a global industry, it is important to take a macro approach and analyze companies across many businesses within the industry. Just think about the logistical and price mechanisms that allow us every day to comfortably supply our vehicles with gasoline or petrol at the local service station. There are many steps along the way and many different companies involved in delivering that gallon or liter of fuel to the nearby service station.

The book is organized into eight chapters:

- Chapter one provides an overview of oil & gas as commodities as well as the industry, introduces the different sectors within the oil & gas industry as well provides an overview of the oil & gas value chain, with a particular emphasis on Refining & Marketing.
- Chapters two provides an overview of financial analysis concepts, processes as well incorporates step-by-step guidance on how to calculate and better understand widely used metrics in financial analysis.
- Chapters three & four introduce the petroleum refining & marketing businesses, explain key processes, their value chain as well takes a deep dive into how each business metrics and key performance indicators are calculated and why they matter.
- Chapter five features three North American-based Refining & Marketing companies with diverse business models and assets.
- Chapter six covers two companies from the Europe, Russia and CIS region.
- Chapter seven covers two companies from the growing Asian & the Middle Eastern markets.
- Chapter eight introduces one company based in Latin America.

Join us on this learning journey as we discover great companies in the exciting petroleum refining & marketing sectors of the oil & gas industry.

Again, thank you for selecting this book!

Alfonso Colombano
Houston, Texas, USA
July 2017

Chapter I – Introduction

"Great works are performed not by strength but by perseverance."   *Samuel Johnson*

The Petroleum Refining & Marketing industry is a fascinating business to be in. Just think about how often we are positively impacted every day by the products we use from the refining & marketing industry. From commuting to work, taking a trip from one corner of the world to the other, to life saving plastics, the refining & marketing industry enriches and improves our lives. To produce a barrel of oil, transport it to a refinery and refine it into gasoline, requires the immense efforts of several diverse upstream, midstream and downstream companies involved in the process.

Audience

Many readers from different backgrounds can benefit from reading this book. This book discusses the importance of the downstream sector of the oil & gas industry and how key this oil & gas sector is. Among some of the audience members who can benefit from reading this book:

- University students interested in the petroleum refining & marketing industry.
- Financial analysts working in or outside the petroleum refining & marketing industry looking to broaden their business knowledge.
- Experienced employees in the oil & gas industry who may have worked in upstream but are interested in learning about the downstream business from an *end-to-end* perspective.
- Analysts or investors looking to learn more about petroleum refining & marketing companies around the globe.
- Anybody interested in expanding their general knowledge about the petroleum refining & marketing industry.

What readers will learn from this book

It is anticipated readers will benefit from learning, at a minimum, the following items from reading this book:

- Understand the different businesses within the petroleum refining & marketing industry, their business cycles, unique opportunities and challenges.
- Understand how financial and operational metrics for companies inside and outside the petroleum refining & marketing industry are calculated and understand their importance.
- Get to know the different companies in the industry, from both an international and U.S. perspective.

- Gain awareness of what different businesses companies in this sector are involved in and where they operate.

Why this book?

As one of the most complex industries in the world, this book provides readers with an in-depth coverage of companies that operate businesses within the downstream oil & gas industry.

Let's begin this journey by defining what oil, natural gas and natural gas liquids (NGL) are, then define what upstream, midstream and downstream are, then follow these definitions by introducing the reader to the three distinct sectors within the larger oil & gas industry, and discuss the different types of companies involved in the petroleum refining & marketing sector.

What is oil?

Oil is a hydrocarbon, typically referred to as *crude oil* in the industry. Oil is a *mixture* of *hydrogen* and *carbon*[1] *primarily* found in liquid state at atmospheric conditions. The hydrocarbon mixtures found in crude oil range in properties, such as boiling points, number of carbon and hydrogen atoms in each molecule, color, viscosity and many other properties. The following paragraph succinctly summarizes what crude oil is:

> *Oil is a fossil fuel. Most of the oil extracted today has been formed from prehistoric organisms whose remains settled at the bottoms of oceans and lakes millions of years ago. As layers of sediment covered them, the pressure on them increased which in turn increased the temperature. This process changed their chemical composition, eventually transforming them into oil*[2].

In the world oil is *primarily* measured in terms of standard U.S. oil barrels at standard conditions. What are these so-called "standard conditions"? Standard conditions are usually defined as volumes measured at a standard temperature of 60 degrees Fahrenheit and standard pressure of 14.65 pounds per square inch absolute (psia)[3]. One barrel of oil is equivalent to 42 U.S. gallons or about 159 liters.

---

[1] Thus why oil is called a *hydrocarbon*

[2] http://www.edfenergy.com/energyfuture/oil

[3] Pounds-per-Square-Inch Absolute, a measure of pressure. For more information, please visit http://ww2010.atmos.uiuc.edu/%28Gh%29/guides/mtr/fw/prs/def.rxml

Composition of Crude Oil

Crude oil is a complex mixture of hydrocarbons. The composition of crude oil varies widely depending on *where* and *how* the crude oil was formed[4]. One crude oil produced from one field is not the same as other crude oil produced from another field[5]. Even among different wells within the *same* field, the composition of crude oil can vary. An additional dynamic is that the composition of crude oil from the *same field* can vary as production declines and wells are drilled within zones in the same field[6].

Below is the typical chemical composition, by weight, of sample crude oils[7]:

- Carbon 84-87%
- Hydrogen 11-14%
- Sulfur 0.06-2%
- Nitrogen 0.1-2%
- Oxygen 0.1-2%

Crude oil can contain other molecules, but the above are the most typical. For a more detailed discussion around crude composition, please refer to chapter three. As mentioned previously, the two most important elements of crude oil, natural gas and natural gas liquids are *hydrogen* and *carbon*, which is the reason these commodities are collectively referred to as *hydrocarbons*[8].

What is density?

Density is a measurement of how solid something is. Density is the *mass* or *weight* of a substance *per unit of volume*. Recall for a moment from your science lab days that every substance has "space" in and around the solids at the molecular level. With that in mind, *density* might be said to be the proportion of the full space *relative* to the empty spaces in any solid, liquid or gas substance.

Density can be defined as a formula of mass *divided* by volume, i.e. grams per liter or pounds per gallon. For example, water at atmospheric pressure

---

[4] http://chemistry.about.com/od/geochemistry/a/Chemical-Composition-Of-Petroleum.htm
[5] Field: An accumulation, pool, or group of pools of hydrocarbons or other mineral resources in the subsurface. A field usually has several or many thousand wells. Source: http://www.glossary.oilfield.slb.com/Terms/f/field.aspx
[6] http://petrowiki.org/Oil_fluid_characteristics
[7] Calculated using crude essays from ExxonMobil. http://corporate.exxonmobil.com/en/company/worldwide-operations/crude-oils/assays
[8] Nontechnical Guide to Petroleum Geology, Exploration, Drilling and Production, 2nd Edition, PennWell books, page 3

has a density of 1 gram per milliliter[9]. In other words, 1 liter of water will have a weight of 1,000 grams or 1 kilo.

In general terms, the more carbon atoms a hydrocarbon molecule has the more *mass* or *weight* it will have. For example, pentane[10], which contains 5 carbon atoms, has more *weight* or *mass* for the *same* unit of volume than methane[11] which only has 1 carbon atom.

In the oil & gas industry, the most widely used measure of density is the American Petroleum Institute gravity rating or simply known as the API gravity[12].   The general API gravity formula is presented here, with G representing the *specific gravity* of the liquid being measured:

$$Degrees\ API = \frac{141.5}{G} - 131.5$$

For example, water, which has a specific gravity of 1.0, will yield an API gravity of 10 degrees:

$$10\ Degrees\ API = \frac{141.5}{1.0} - 131.5$$

It can then be concluded that any crude oil with an API gravity *lower* than 10 degrees would *sink* in water, due to the fact that this crude oil is *heavier* or more *dense* than water. Conversely, any crude oil with an API *higher* than 10 degrees would *float* on top of water, since this crude oil is *lighter* than water.

Types of Crude Oils

Crude oils from many different wells or fields differ in quality from each other, with oil being categorized in terms of density and other properties. Lighter crude oils have a higher API gravity scale[13], while heavier crude oils have a lower API gravity. Light crude oils have hydrocarbon molecules with *less* carbon atoms (thus why they are called *light* crude oils) and are easier to produce and refine than heavy oils. Typically[14], lighter oils tend to command a higher price in the market than heavier oils[15].

---

[9] Water has a mass of 1 gram per milliliter at 4 degrees Celsius (39.2 F)

[10] Pentane, from the Greek word *penta* or five, and ane from *alkanes*

[11] Methane is the simplest hydrocarbon and has only 1 carbon atom and 4 hydrogen atoms, thus the formula CH_4

[12] http://www.sizes.com/units/hydrometer_api.htm

[13] For more information, please visit: http://total.com/en/energies-expertise/oil-gas/exploration-production/strategic-sectors/eho/challenges/presentation

[14] See here for more information: http://www.baytexenergy.com/operations/marketing/benchmark-heavy-oil-prices.cfm

[15] See more information: http://economicdashboard.alberta.ca/OilPrice

Lighter crude oils are generally considered as those having an API scale higher than 35, medium crude oils are categorized as those having an API gravity between 27 and 34 degrees, while heavy oils typically have an API lower than 27. As discussed before, the API gravity of water is 10, so any crude oil with an API lower than 10 would sink in water, which is the case with crude oils that are considered *extra-heavy*, such as Venezuelan bitumen and Canadian Tar Sands[16]. The following table provides a sample of different crude oil types, their API gravities and corresponding sulfur content[17]:

Common Crude Oil Type	API Gravity (degrees)	Sulfur Content (%)
Light Sweet	35-50+	Less than 0.3%
Light Sour	35-40	Less than 1.1%
Medium Sour	27-34	Greater than 1.1%
Heavy Sweet	Less than 27	Less than 1.1%
Heavy Sour	Less than 27	Greater than 1.1%
Extra-Heavy	Less than 10	Greater than 1.1%

There are many quality factors affecting crude oil, but one of the most important quality factors besides gravity is sulfur content. The higher the sulfur content, the *less valuable* the crude oil tends to be since sulfur is a highly corrosive element[18]. In order for a refinery to process high sulfur crude oils, the refinery requires highly specialized desulfurization process equipment as well as more expensive metallurgy[19]. The *higher* the sulfur content the more *"sour"* the crude oil is. Correspondingly, the *lower* the sulfur content the more *"sweet"* the crude oil is.

Why are crude oils classified as "sour" or "sweet"? This practice actually originated in the 19th century, when drillers would literally *taste* or *smell* the different crude oils and thus described the different types of oils as having a *sweet* or *sour* smell or taste[20].

---

[16] https://thetyee.ca/News/2013/05/23/Bitumen-Does-Not-Float/
[17] Energy Information Agency: http://www.eia.gov/analysis/petroleum/crudetypes/
[18] Sulfur causes high-temperature sulfidation of metals, and it combines with other elements to form aggressive compounds, such as various sulfites and sulfates, requiring special types of steel. For more information:
http://www.calbrite.com/resources/role%20of%20ss%20in%20petroleum%20refining.pdf
[19] http://smt.sandvik.com/en/applications/oil-and-gas-downstream/
[20] http://setxind.com/upstream/in-depth-look-at-crude-oil/

The following table provides a sample of different crude oils from around the world, their API gravity and corresponding sulfur content[21]:

Country/Area	Crude Oil Name	API Gravity (degrees)	Sulfur Content (%)
Algeria	Sahara Blend	43.6	0.0725
United States	West Texas Intermediate (WTI)	39.6	0.3
Mexico	Olmeca	38.7	0.891
North Sea	Brent	38.6	0.38
Libya	Es Sider	36.5	0.402
Australia	Barrow	36.1	0.05
Canada	Hibernia	34.4	0.406
Nigeria	Bonny Light	33.9	0.529
Russia	YK Blend	33.7	0.756
Saudi Arabia	Arab Light	33	1.98
United States	Light Louisiana Sweet (LLS)	32.9	0.35
United States	West Texas Sour (WTS)	32.8	1.98
Caspian Region	Urals	31.8	1.24
Kuwait	Kuwait	30.1	2.8
Iraq	Basrah Light	30.1	2.8
United States	Mars Blend	28.6	2.02
Saudi Arabia	Arab Heavy	27.7	2.99
Venezuela	Hamaca	25.5	1.6
Ecuador	Oriente	24	1.5
Canada	Albian Muskey River Heavy	21.6	3.81
Canada	Western Canadian Select (WCS)	20.3	3.43

Crude Oil Value Chain

In the crude oil value chain, there are several high level components:

- Oil and condensate wells, which produce crude oil and condensate.
- Separation facilities, which separate oil, gas & water as they are produced.
- Crude oil is transported by trucks or gathering pipelines *from* producing wells *to* receipt points such as *terminals[22]*, long-haul pipelines receipt points, or even transportation vessels or ships that transport this valuable crude oil and condensate where it is most valued.
- Crude oil is received at refineries and transformed into *higher value* products such as gasoline, diesel, jet fuel, lubricants and other useful products.

---

[21] http://www.caplinepipeline.com/PDF/report1.pdf
[22] A facility used to receive, store or deliver crude oil and petroleum products

- Refined petroleum products are transported to products terminals, pipelines or even transportation vessels or ships around the world.
- Refined petroleum products are delivered to service stations around the world and are subsequently sold to consumers.
- Petroleum products are used around the world for a variety of needs, such as transportation, heating, power generation, plastics, manufacturing and many other uses.

Crude Oil Marketing

Crude oil is a global commodity bought and sold in many different parts of the world. Crude oil is the most actively traded commodity in the world[23]. Crude oil marketing generally entails purchasing crude oil from producers at the wellhead and other trading locations and then selling it to a refinery or to other buyers[24]. Crude oil marketing is covered in more detail in chapter four.

What is crude oil used for?

Crude oil is the most valuable commodity in the world, primarily due to its versatility in the many different products that can be produced from this raw material. Typically, for 1 barrel or 42 gallons of crude oil[25], the following products can be refined:

Typical Product	Gallons	Percent of total
Gasoline	19.74	47%
Diesel Fuel & Heating Oil	9.66	23%
Other Products	7.56	18%
Jet Fuel	4.2	10%
Liquefied Petroleum Gas or LPGs	1.68	4%
Asphalt	1.26	3%
Total	**44.1**	**105%**

As can be seen, the total volume of products, 44.1 gallons, does not equal the original 42 gallons of crude oil processed. Why is this? This is due to what is called in the industry *refined volume gain or loss* or processing gain. One barrel of gasoline or jet fuel has different *density* than one barrel of crude oil. In other words, since crude oil is *heavier* than gasoline, a barrel of crude oil and gasoline will occupy the *same volume*, but the crude oil barrel will "*weigh*

[23]http://commodities.about.com/od/researchcommodities/a/most-liquid-commodity-markets.htm
[24] http://www.sunocologistics.com/Customers/Business-Lines/Crude-Oil-Acquisition-and-Marketing/173/
[25] http://www.eia.gov/dnav/pet/pet_pnp_pct_dc_nus_pct_a.htm

more" or have more hydrocarbon molecules than gasoline. Refined volume gain can be described as follows:

> "*Refined volume gain: The volumetric amount by which total output is greater than input for a given period of time. This difference is due to the processing of crude oil into products which, in total, have a lower specific gravity than the crude oil processed.*"[26]

Refined volume gain and loss is covered in more detail in chapter three of this book.

The most important and largest category of refined products consumed in terms of volumes are transportation fuels. Transportation fuels are primarily gasoline (petrol in the UK and Australia), diesel fuel and jet fuel. The primary source for these transportation fuels is crude oil[27]. For the foreseeable future, crude oil derived transportation fuels are expected to continue to dominate the transportation sector. Liquid fuels provide the *reliability, availability and high energy density*[28] that consumers around the globe have come to expect.

Many other products are refined from crude oil, and include asphalt, petroleum coke and Liquefied Petroleum Gas or LPGs. These other products generally tend to be of *lower* value and economic impact than transportation fuels. As such, refiners in general try to *maximize* yields of transportation fuels while seeking to *minimize*, to the extent possible, the production of products such as asphalt, petroleum coke, LPGs and others.

Crude oils are also extensively used in petrochemicals, with some refined products from crude oil competing for market share with lower cost natural gas liquids feedstocks[29]. Natural Gas Liquids, an important feedstock[30] in the petrochemicals industry, are discussed in more detail later in the chapter.

What is natural gas?

Natural gas, like crude oil, is a mixture of hydrocarbons, which is primarily found in *gaseous* or *vapor state* at atmospheric conditions. The main

---

[26] http://www.eia.gov/dnav/pet/TblDefs/pet_pnp_refp2_tbldef2.asp
[27] Certain refined products can also be obtained from processes such as biodiesel or Gas-to-Liquids technologies, but by far the biggest source of transportation fuels is crude oil.
[28] High energy density allows more energy to "fit in" in the same amount of volume.
[29] For more information, see http://www.trefis.com/stock/xom/articles/244267/exxon-mobil-breaks-ground-on-new-chemical-plant-in-the-u-s-to-tap-lower-natural-gas-prices/2014-06-23
[30] Feedstock: Raw material (input) fed into a process for conversion into something different (output). Crude oil derived products and NGLs are both feedstocks into petrochemical industries

component of natural gas is methane (CH_4)[31]. Methane or CH_4 is the main constituent of *pipeline quality gas*, representing upwards of 90% of natural gas. In addition, natural gas also contains what are called natural gas liquids or NGLs. NGLs are ethane, propane, butanes and pentanes plus[32].

Natural gas, in its gaseous state, can only be easily transported by pipelines. Natural gas located in remote regions (in industry terms "stranded gas") can be converted into Liquefied Natural Gas or LNG. In general terms, an LNG plant chills down natural gas to a temperature of *negative* 260 degrees Fahrenheit. By cooling down the natural gas to cryogenic temperatures, this chilling process allows the gas to occupy significantly less volume, causing natural gas to *decrease* in volume by a factor of $1/600$[33]. This natural gas, now in liquid form and occupying *much less* volume, can then be economically shipped in refrigerated vessels hundred or even thousands of miles away from its original production site.

The other alternative method to transport natural gas is to compress the gas to 3,000-4,000 pounds per square inch (psi). By compressing the gas at these high pressures, the compressed natural gas or CNG occupies less than 1% of the original volume. By doing this natural gas can be more easily transported since it occupies less space. CNG is also used as a transportation fuel, particularly in fleet vehicles. Natural gas has unique impacts to refining processes, in particularly in terms of providing associated NGL feedstocks, hydrogen production and energy to various refining processes. The impact of natural gas is covered in more detail in chapter three.

What are NGLs?

NGLs or natural gas liquids are valuable hydrocarbons that are usually found associated with natural gas or that can be refined from crude oil. Why are they called NGLs then? They're called NGLs because these hydrocarbons are found in gaseous state at *atmospheric conditions* but can be converted into a liquid by either *increasing* the pressure, *decreasing* the temperature or *both*. NGLs are usually more valuable than pipeline quality natural gas and thus provide hydrocarbon producers an additional incentive to extract natural gas liquids from the natural gas stream. The other reason NGLs need to be extracted from the natural gas stream is the requirement to comply with necessary natural gas pipeline specifications regarding excess NGL content.

---

[31] Natural gas composition depends on the particular well and field where it was produced
[32] Pentanes+ also go by different names, such as natural gasoline, condensates, and other names. For more information: http://www.eia.gov/petroleum/workshop/ngl/pdf/definitions061413.pdf
[33] http://www.plumenergy.com/liquefied-natural-gas-faq/

The following table from the Energy Information Agency describes the different end-uses for NGLs[34]:

Natural Gas Liquid	Chemical Formula	Applications	End Use Products	Primary Sectors
Ethane	C_2H_6	Ethylene for plastics production; petrochemical feedstock	Plastic bags; plastics; anti-freeze; detergent	Industrial
Propane	C_3H_8	Residential and commercial heating; cooking fuel; petrochemical feedstock	Home heating; small stoves and barbeques; Liquefied Petroleum Gas (LPG)	Industrial, Residential & Commercial
Butane	C_4H_{10}	Petrochemical feedstock; blending with propane or gasoline	Synthetic rubber for tires, LPG, lighter fuel	Industrial & Transportation
Iso-butane	C_4H_{10}	Refinery feedstock; petrochemical feedstock	Alkylate for gasoline to increase octane rating; aerosols; refrigerant	Transportation
Pentane	C_5H_{12}	Natural gasoline; blowing agent for polystyrene foam	Gasoline; polystyrene; solvent	Transportation
Pentane Plus	Mix of C_5H_{12} and heavier	Blending with vehicle fuel; exported for bitumen production in oil sands	Gasoline; ethanol blends; oil sands production	Transportation

On the table above, C indicates the number of carbon atoms while H indicates the number of hydrogen atoms, i.e. Ethane contains 2 carbon atoms and 6 hydrogen atoms. NGL's are covered in more detail in chapter 4 under the NGL trading section.

What is Upstream, Midstream and Downstream?

The oil & gas industry is traditionally divided into three *distinct* sectors, upstream, midstream and downstream:

- Upstream, also known as Exploration & Production is primarily focused on *finding, exploring, developing* and *producing* crude oil and natural gas. These hydrocarbons are then gathered and processed from the producing well and typically sold to Midstream or Downstream companies at a terminal, gas plant, refinery or other delivery points.

---

[34] http://www.eia.gov/todayinenergy/detail.cfm?id=5930

- Midstream is *typically* involved in transportation and intermediate processing of raw petroleum products, such as oil and natural gas, as well as finished petroleum products and other hydrocarbons.. Conventionally, midstream has encompassed *gathering, treating, processing* of natural gas, long haul transportation of oil, natural gas, natural gas liquids, refined products and chemicals, as well as terminal assets, LNG, LPG and crude oil shipping, fractionation/separation of NGLs and other areas. For several companies, especially the large integrated companies, their "midstream assets" are usually grouped or reported as Downstream instead of as a standalone midstream segment.

- Downstream, also known as Refining & Marketing, is principally involved in *refining, marketing, transporting* and *retailing* finished petroleum products, such as gasoline, diesel, jet fuel, natural gas and other petroleum products to wholesale and end customers. The downstream sector in the oil & gas industry also encompasses other products such as petrochemicals, lubricants, and other businesses.

The following table provides a summary of the activities, business models, price risk, on-going capex requirements, typical levels of profitability as well as other risks that are found in these three oil & gas sectors: [35]:

	Upstream	Midstream	Downstream
Main Activity	Finding, exploring and producing hydrocarbons	Transporting and performing intermediate processing of hydrocarbons	Refining and delivering petroleum and petrochemical products
Business Model	Mining	"Toll-Road"	Manufacturing/Retail
Price Risk	Dependent on high commodity prices	Relatively low price risk. Primarily Fee-based earnings	Margin business, benefits from a high spread between *inputs* (crude oil) and *outputs* (petroleum products)
On-going CAPEX	High level, both on-going and initial capex	Initial capex high, moderate to low on-going capex	Initial capex high, moderate to low on-going capex
Profit Margin	High to Medium	Medium to Low	Low to Medium
Cash Flow Stability	Highly Volatile	Most stable	Medium to High Volatility
Other Risks	Geopolitical & Environmental	Environmental	Environmental

---

[35] From Oil & Gas Company Analysis: Upstream, Midstream & Downstream by Alfonso Colombano

Downstream Overview

The downstream business can generally be said to encompass the following activities:

- Refining of raw crude oil and other feedstocks into petroleum products, such as gasoline, diesel, jet fuel, asphalt and others.
- Wholesale and retail marketing of refined petroleum products
- Shipping of crude oil and refined petroleum products.
- Manufacturing of base oils, lubricants, and other specialty *petroleum-based* products.
- Manufacturing, distribution and marketing of petrochemicals.
- Distributing natural gas, although this activity is also classified by some companies classified as midstream.

Historically both Refining & Marketing have been treated, for external reporting purposes, as a single combined segment, usually with the acronym "R&M" to denote all downstream activities. This is particularly applicable in the large integrated companies, such as ExxonMobil, Chevron, Shell and others[36].

Refining

Refining is usually the largest business within downstream oil & gas in terms of earnings, capital employed and volume. Refining is covered in more detail in chapter three.

Marketing

The marketing operations in downstream encompass both *wholesale* and *retail* marketing. Retail marketing is one of the areas in the oil & gas industry that has a large end-customer presence and most consumers have come to associate intrinsically with the industry. Marketing is covered in more detail in chapter four.

Petrochemicals

Apart from Refining & Marketing, the petrochemicals business is the largest business in the downstream oil & gas industry. This business is primarily engaged in the *production*, *distribution* and *sale* of a wide variety of petrochemicals, from *commodity* chemicals such as polyethylene,

---

[36] One key exception of this trend, has been Total, which has a substantial petroleum marketing business in Europe and Africa.

polypropylene to more specialized chemicals used, such as drilling fluids, odorants to natural gas to many others[37].

Petrochemicals are essential in today's economy and are used in manufacturing a wide variety of consumer goods such as from cellphones, auto interiors, to carpeting, to food packaging, to electronics. Simply put plastics derived from oil & gas are widely used everywhere. The table below provides an overview of the six basic feedstocks and associated products:[38]

Basic Petrochemical	Sample Applications & Products
Ethylene	Plastic bags, bearings, clothing, credit cards, detergent bottles, electrical cables, engine coolant, film, gears, plastic bottles, pipes, polyester, signs, food packaging, adhesives and other components
Propylene	Adhesives, appliance parts, automotive plastic components, carpets, coating, cups, containers, disposable diapers, fixtures, housewares, furniture, insulation, paints, textiles, electric & electronic appliances, construction materials, cosmetics, marine industry parts, film and sheets, paints & coatings, inks
Butadiene	Automotive bumper bars, automotive components, computer keyboard keys, golf club heads, tires, automotive hoses and belts, latex paints, luggage, kitchen appliance parts, rubber, toys
Benzene	Automotive headlamps, cutlery, insulation, computer cases, instrument strings, nylons, phones, packaging material, rope, safety glasses, sunglasses, tents,
Toluene	Boat parts, fabrics, food packaging, furniture, nylon, textiles, clothing, varnish
Xylene	Automotive applications, beverage bottles, carpets, fabrics, electronic, solvents, sportswear, textiles

The U.S. petrochemical industry, thanks largely to rising U.S. production of hydrocarbons, has experienced considerable growth over the past several years[39]. Thanks to *low cost* feedstocks, such as ethane and natural gas, the U.S. petrochemical industry has one of the *lowest* cost curves for several petrochemical products, right behind Middle East producers. This feedstock advantage is expected to continue over the next couple of years [40]. For a more in-depth analysis of the petrochemical industry, we recommend referencing William L. Leffler's *Petrochemicals in Nontechnical Language*, published by PennWell books.

Other businesses in Downstream

Downstream is a sector that is probably as difficult to define as Midstream, since it encompasses so many other areas other than Refining & Marketing.

---

[37] http://www.cpchem.com/bl/specchem/en-us/Pages/default.aspx?Redirect=1
[38] https://www.afpm.org/uploadedFiles/Content/Our_Members/What_We_Make/Petrochemicals/petrochemical-infographic.pdf
[39] http://www.pwc.com/en_US/us/industrial-products/publications/assets/pwc-shale-gas-chemicals-industry-potential.pdf
[40] http://www.chemweek.com/lab/Petrochemicals-Huge-midstream-investments-underpin-rebirth-of-US-industry_59746.html

Many of these other businesses are usually embedded in downstream companies' earnings. Especially for the larger companies, there is usually not very granular reporting for some of these businesses. These businesses tend to be high margin businesses but usually have relatively low volumes *versus* the bigger R&M business. There are other businesses that are usually not considered "downstream" in the traditional integrated oil & gas company sense, but truly serve a *downstream* function like natural gas local distribution companies.

Although Refining, Marketing & Petrochemicals are the biggest components in downstream, there are other businesses that are part of this sector:

- Lubricants business: This business is primarily engaged in the *production* and *marketing* of lubricants, motors oils, greases and other products with applications in various industries such as ground, air, maritime transportation, industrial machinery and others.

- Specialty Products & Chemicals: This business usually includes products such as flow improvers and specialty chemicals used in very discrete applications across a varied group of industries.

- Natural Gas Local Distribution Companies (LDCs): These companies can also be classified as *utilities* due to the fact that most of these companies are regulated across the globe and the U.S. This industry is also typically classified as Midstream operations or activities, primarily due to their business model being more similar to a *"toll road"* or more like a typical regulated utility in the United States and many other countries.

- Crude and Refined Products Supply & Distribution companies: These companies can be classified as well as "midstream" companies. These companies transport crude oil and refined products primarily via trucks or barges. This transportation can take place by collecting or "hauling" crude oil from different producing oil wells across the country to be delivered to crude *terminals*[41] or by transporting refined products from products terminals to service stations across the country. These companies tend to be relatively small and are usually not publicly traded. These companies are critical to the oil & gas sector, since they assure a continuous product outflow and provide market access to small

---

[41] A terminal is a receipt, delivery and storage point for commodities.

producing oil wells, particularly in the United States[42]. These companies are critical since they deliver petroleum products to service stations around the globe and are basically the last step in the value chain.

- Supply & Trading operations provide a critical link for companies' downstream operations by supplying refineries with the lowest cost of crude while marketing refined products at the highest value possible. Supply & Trading operations are covered in more detail in chapter four.

Oil & Gas Companies in this book

The companies covered in this book represent a wide variety of downstream players, including:

- Downstream operations of National Oil Companies (NOCs)
- Downstream operations of Integrated Oil Companies, also known as, *International Oil Companies* (IOCs)
- Independent Downstream, also known as *Pure Play Refining Companies*

National Oil Companies (NOCs)

National Oil Companies or NOC's are the biggest participants in the upstream oil & gas industry in terms of hydrocarbon *reserves* and *production*. Although NOC's have become bigger players in downstream, NOC's downstream operations are not as *developed* and *significant* for their total operations as their upstream assets. Beginning in the 1950's and continuing through the 1980's the creation of NOC's became more prevalent. One of the main drivers for the creation of NOC's was that host governments wanted to increase participation in the oil & gas industry in their own countries. NOCs can be further divided into two categories:

- Non-Publicly Traded National Oil Companies
- Publicly Traded National Oil Companies

Non-Publicly Traded National Oil Companies

As their name implies, these companies are fully owned and/or controlled by their respective governments and do not allow individual investors to own shares in these companies. These companies are among the largest

---

[42] Crude trucks are critical in the US for gathering crude oil from remote wells. For more information: https://fas.org/sgp/crs/misc/R43390.pdf

companies in the world and are key players in the oil & gas markets. They are usually integrated since they have upstream, midstream and downstream operations. Examples of some of the companies in this category are PDVSA, PEMEX, and Qatar Petroleum.

Publicly Traded National Oil Companies

Publicly traded NOC's have a substantial ownership or control by their respective host governments, but allow individual investors to own shares in these companies. These publicly traded NOCs are usually listed on major stock indexes around the world such as New York Stock Exchange (NYSE), London Stock Exchange (LSE), Moscow Stock Exchange (MOEX), among others. These companies allow investors to *buy* and *sell* shares of their companies on exchanges around the world through what are known as American Depositary Receipts or ADRs[43] (also known as Global Depositary Receipts or GDRs). These companies may also invest beyond their home countries and are expected to compete more and more in the future with the International Oil Companies (IOC). Examples of these companies are Statoil, Petrobras and Petrochina.

Integrated Companies / International Oil Companies (IOC) / Super-majors

These are the companies most traditionally associated with the oil & gas industry such as ExxonMobil, Shell, BP and Total. These companies are *integrated* in the sense that they have operations in all three sectors, *upstream*, *midstream* and *downstream* around the globe.

Downstream operations of an Integrated Oil Company

Some of the world's largest downstream companies are integrated companies, such as ExxonMobil, with industry leading downstream and chemicals operations. IOC's own extensive assets in the downstream sector of the industry and may even process crude oil produced from the company's upstream assets. The very fact that IOC's own and operate downstream assets is what makes these companies have a *built-in* ability to withstand different economic environments. Downstream is a cyclical sector and usually does very well when *input costs are low* and *refined products prices are high*, thus making margins *high*.

---

[43] For more information, please visit http://www.investopedia.com/terms/a/adr.asp

IOC's own or operate downstream assets such as:

- Large, *complex* integrated refineries located in many established refining centers, such as the U.S. Gulf Coast, Europe, Asia and the Middle East.
- Product pipelines and associated terminal operations.
- Large petrochemical operations that produce both commodity petrochemicals as well specialty petrochemicals. These petrochemical operations are usually integrated with the company's refining assets, benefiting from feedstocks produced as by-products of their refining operations and sharing utilities. Companies such as ExxonMobil benefit from a very close petrochemical and refining integration, particularly in their U.S. Gulf Coast and European refineries[44].
- Wholesale and retail marketing operations both in the U.S. and Internationally.
- Production of base oils, lubricants, motor oil, additives and other industrial and consumer products.

Independent Downstream / Pure Play Refining Companies

Independent Downstream, also known as pure play Refining companies, derive most of their revenues, earnings and cash flow *primarily* from refining crude oil into valuable products such as motor gasoline, diesel, jet fuel and other products. These companies may also have wholesale and retail fuel marketing assets. Historically, the refining & marketing sector of the industry has been quite volatile, and a few of these companies have pursued growing their other businesses, such as chemicals, midstream or lubricants. As an example, two companies, Marathon Petroleum and Phillips 66, joined this category of companies, which were spun-off from their integrated companies in 2011 and 2012 respectively.

Independent downstream companies, as the name implies, have operations in the downstream sector of the oil & gas industry. Downstream companies can be generally described as having the following characteristics:

- Refining assets represent their largest operations in terms of capital employed, revenues and earnings.

---

[44] 2016 ExxonMobil Financial & Operating Review, page 10

- Earnings are subject to the cyclicality of refining margins and can vary substantially[45] from *quarter* to *quarter* and *year* to *year*.

- Generate substantial amounts of total cash from operations from their refining operations and may re-invest this cash into other *higher-margin* areas, such as midstream, marketing or chemicals[46].

- Many refining companies have leveraged their transportation assets such as pipelines, terminals and fractionators[47] into forming Master Limited Partnerships or MLPs to capture higher market valuations[48].

- Devote a *significant percent* of cash flow from operations to shareholder distributions (both dividends and share repurchases).

Units and Conversion

The oil & gas industry is unique in that it *largely* uses roman numerals to denote large figures. The Roman numeral "M" stands for 1 *thousand* and *not* 1 million. The roman numerals MM would stand for one thousand *times* one thousand or simply 1 *million*. Consistent with industry practice, throughout this book the following units are used:

- M is equal to one *thousand* or 1,000.
- MM is equal to one *million* or 1,000,000.
- 1 billion is equal to one *thousand* million or 1,000,000,000.
- 1 trillion is equal to one *million* million or 1,000,000,000,000.

Comma, Periods and Currency

Consistent with industry practice, the comma sign is used to facilitate reading of quantities greater than one thousand. Similarly, the period is used to denote a decimal point:

- For example one million U.S. Dollars is written as $1,000,000.00.
- To denote quantities, the convention to be used is $MM for million U.S. dollars. For example, one-hundred fifty million U.S. Dollars is denoted as $150MM.
- For quantities larger than one thousand million dollars, billions are used. For example, 1,000,000,000.00 is equal to $1 billion.

[45]www.eia.doe.gov/pub/oil_gas/petroleum/analysis_publications/petroleum_issues_trends_1996/CHA PTER7.PDF

[46] http://investor.phillips66.com/investors/news/news-release-details/2014/Phillips-66-Announces-2015-Capital-Program/default.aspx

[47] http://fuelfix.com/blog/2016/02/18/phillips-66-sells-stake-in-new-projects-to-its-mlp/

[48] http://www.marathonpetroleum.com/News/News_Releases/Press_Release/?id=1655082

Throughout this book, the currency used is the United States Dollar or "U.S. Dollar". The symbol used to denote U.S. Dollars is the $ symbol. Financial statements published in a currency other than the U.S. Dollar have been converted using their respective exchange rates. Currencies reviewed in this book are follows, with average rates in 2016 of:

Currency	Symbol	Exchange Rate (Currency units per 1USD)	Company
European Union Euro	EUR	0.90	Total SA
Chinese Yuan	CNY	6.94	Petrochina
Russian Ruble	RUB	67.03	Rosneft
Saudi Riyal	SAR	3.75	Saudi Aramco
Brazilian Real	BRL	3.49	Petrobras

Whenever foreign companies publish statements already translated to US dollars, these statements would be presented in their USD conversion instead of using the published rate since this reflects the underlying average USD rate instead of a year-end rate.

Crude oil + Liquids Units

In the oil & gas industry, crude oil, natural gas liquids, bitumen, tar sands and other liquid hydrocarbons, are measured in U.S. barrels.

Consistent with industry practices, throughout this book, barrels are used as the unit of measure for oil and other liquid hydrocarbons. One (1) U.S. barrel is defined as 42 U.S. gallons at atmospheric conditions, with atmospheric conditions being defined as 60 degrees Fahrenheit and pressures of 14.65 pounds of per 1 square inch or psi. The abbreviation for a U.S. barrel is "bbl" and this abbreviation is also used throughout this book.

Production and refining of liquid hydrocarbons is usually measured in barrels per day. The following are the measurement and acronym conventions used throughout the book:

- BPD = Barrels of oil or liquids *per* day.
- MBPD = *Thousand* barrels of oil or liquids *per* day.
- MMBPD = *Million* barrels of oil or liquids *per* day.

For measuring liquid hydrocarbon reserves, the following convention is used:

- Mbbl = *Thousand* barrels of oil or liquids reserves.
- MMbbl = *Million* barrels of oil or liquids reserves.
- For units concerning reserves larger than 1,000MMbbl, billion is used, which equals to 1,000,000,000 or one *thousand* millions as in 10 billion barrels.

Metric units

Certain companies around the world, for example those headquartered in China, Norway and Russia, measure oil by *weight* instead of *volume*, denoted in terms of one metric ton. One metric ton of oil is equivalent to about 7.33 barrels of oil[49], but the exact measure depends on the *type* of oil and *gravity*. A metric ton is a measure of *mass* or *weight*, while a barrel is a measure of *volume;* therefore, the conversion is dependent on the *density* of the particular oil or product being measured and will vary from year to year. Throughout the book, *except where noted*, flows of crude oil and products measured by companies originally in metric tons have been converted using the 7.33 barrels to 1 metric ton factor. Natural gas, whenever is measured in cubic meters, *except where noted*, has been converted in the book using a 35.31 cubic feet to cubic meter factor.

The following table provides commonly used conversion factors for a wide variety of hydrocarbons[50]:

Product	From Unit	Operation	Factor	Unit to
Crude Oil	Metric ton	Multiply by	7.33	Barrels
Gasoline	Metric ton	Multiply by	8.35	Barrels
Kerosene	Metric ton	Multiply by	7.88	Barrels
Gas oil / diesel	Metric ton	Multiply by	7.46	Barrels
Residual fuel oil	Metric ton	Multiply by	6.35	Barrels
Natural Gas	Cubic Meter	Multiply by	35.3	Cubic Feet
Natural Gas	Thousand Cubic Feet	Divide by	6	BOE

---

[49] Conversion factor based on worldwide gravity. BP Statistical Review 2014
[50] BP Statistical Review 2016 conversion factors

Chapter II - Financial Analysis Overview

"In investing, what is comfortable is rarely profitable." - Robert Arnott

Financial Analysis

The field of financial analysis is a fascinating, exciting and value-adding activity. Financial analysis is particularly important in the Oil & Gas industry, where each of the major sectors has a different business cycle, operating parameters, capital requirements and different business drivers. Financial analysis can help with analyzing and understanding companies' business structure, key performance indicators, financial and operational drivers and other areas.

Financial analysis can assist with:

- Better understand a company's underlying business.
- Provide capital allocation criteria for an individual's portfolio or a company's portfolio.
- Compare companies against each other in the same sector, geographical area or among different industries.
- Assist in management's decisions to expand, reduce or sell a business.
- Understand why a company's stock performance is higher than its competitors.

Financial Statements

There are three basic financial statements[51]:

- Income Statement, also known as Profit & Loss Statement, reflects revenues, costs and the perennial *bottom line* or earnings for the company for a specific period of time, usually a month, quarter or year.
- Balance Sheet reflects the company's assets, liabilities and owners' equity.
- Statement of Cash Flows, reports on the underlying activity of the company's cash balance for a *specific period* of time and is subdivided into three sections, the operating section, the investing section and financing section.

Let's review each of the first three financial statements in more detail.

---

[51] https://www.sec.gov/reportspubs/investor-publications/investorpubsbegfinstmtguidehtm.html

Income Statement

An income statement is a report that shows how much revenue a company earned over a specific *time period* (usually for a year or a quarter). An income statement also shows the costs and expenses associated with earning that revenue. The literal "bottom line" of the statement usually shows the company's net earnings or losses.

The following table discusses the major line items found on an income statement:

Major Group	Income Statement Line	Description
Inflows	Sales & Other Operating Revenues	Gross Revenues or Sales of a company, which are the amount a company receives from selling hydrocarbons or transportation services. This amount is a gross amount since it does not include any costs or expenses
Inflows	Equity Earnings	*Proportional* share of earnings of affiliate companies owned by parent company that are *unconsolidated*[52]
Inflows	Other Income	Miscellaneous income not part of general operations
Outflows	Purchased crude oil and Products, also known as Cost of Goods Sold	Costs of feedstocks used in the production of refined products, such as crude oil and other hydrocarbons. Included here are also direct inputs such as natural gas
Outflows	Operating Expenses	Expenses incurred *directly* attributable to the production or manufacturing facilities of a company, for example a refinery's employees' salaries, costs of catalysts, electricity at the refinery and other expenses.
Outflows	Selling, General & Administrative Expenses	*Indirect* costs and corporate costs for support staff, such as the CEO, IT, Finance, HR, Marketing and other groups
Outflows	Depreciation & Amortization	Depreciation is the allocation of costs of fixed assets based on different useful lives for each asset. For example, a company-owned vehicle may have a useful life of 5 years, so if the cost is $20,000, each year, $4,000 will be recorded as depreciation expense. D&A are usually the biggest *non-cash* expenses of a company
Outflows	Other Taxes	Non-Income taxes such as payroll, excise, sales, fuel and other taxes
Outflows	Interest & Debt Expense	Costs of interest and issuing debt on an *accrual accounting* basis, since some interest may be capitalized.
Balance	Income Before Taxes	The balance remaining of taking all inflows or revenues and taking out all costs *before income taxes*
Outflows	Income Tax	Income taxes *accrued* by the company for all jurisdictions that the company operates in. These are different from *cash taxes*.
Balance	Net Income	Balance remaining after all costs have been taken out from revenues, but before the minority interest is taken out[53].
Balance	Net Income Attributable	Balance remaining or *bottom line* that remains after subtracting all costs from revenues and taking out minority interest. Thus it is the net income *available* for *current* to holders of common stock (shareholders)

---

[52] An unconsolidated subsidiary is a company that is owned by a parent company, but whose individual financial statements are not included in the consolidated or combined financial statements of the parent.
[53] Minority interest is covered in more detail further in this chapter

Balance Sheet

A balance sheet provides detailed information about a company's assets, liabilities and shareholders' equity. A balance sheet is also known as the *Statement of Financial Position*.

Assets are things that a company owns that have value. Assets include *tangible* assets like cash, accounts receivables, inventory, investments, physical property, plant & equipment, land and other items as well as *intangible* assets like goodwill, brand recognition and intellectual property. Assets are typically listed on the balance sheet in order of how quickly they can be converted to cash and are divided into current assets (those that can be converted to cash within one year) and non-current assets which are longer than one year.

Liabilities are amounts of money or *obligations* that a company owes to others. Liabilities can include loans from banks, bondholder, accounts payable, payroll, environmental cleanup costs, taxes due, unearned or deferred revenues, legal contingencies and other areas. Liabilities are typically listed based on their due dates, starting with accounts payables which are usually paid within 30 days and other current liabilities which have to be paid within one year and non-current or long-term liabilities which are due longer than one year[54].

Equity is sometimes called capital or net worth. Equity is what is left off after *subtracting* out liabilities from assets, in other words, it is the *residual* or *book value* that is left or owned by the owners or shareholders of a company.

A balance sheet shows a snapshot of a company's assets, liabilities and shareholders' equity at the *end of the* reporting period. It does not show the *flows* into and out of the accounts during the period.

A balance sheet is called a *balance sheet* since it has to balance according to the universal accounting equation:

Assets = Liabilities + Owners' Equity

---

[54] Please note that the for US-GAAP companies, typically current assets and liabilities are displayed first in order while for IFRS companies, non-current assets & liabilities are listed first in order.

Cash Flow Statement

Cash flow statements report a company's inflows and outflows of cash[55]. This is important because a company needs to have enough cash on hand to pay its expenses and purchase assets. While an income statement can provide insights as to whether a company made a *profit*, a cash flow statement can describe *how much cash* the company generated.

A cash flow statement shows changes *over time* rather than *absolute* dollar amounts at a point in time. It uses and reorders the information from a company's balance sheet and income statement.

The bottom line of the cash flow statement shows the net *increase* or *decrease* in cash for the period. Generally, cash flow statements are divided into three main parts. Each part reviews the cash flow from each of the three main areas: operating activities, investing activities and financing activities.

Operating Activities

The first part of a cash flow statement analyzes a company's cash flow from net income or losses. For most companies, this section of the cash flow statement reconciles the *net income* (as shown on the income statement) to the *actual cash* the company received from or used in its operating activities. To do this, it adjusts net income for any non-cash items (such as adding back depreciation expenses) and adjusts for any cash that was *used* or *provided* by other operating assets and liabilities.

Investing Activities

The second part of a cash flow statement shows the cash flow from all investing activities, which generally include purchases or sales of long-term assets, such as property, plant and equipment, as well as investment securities. If a company buys a piece of machinery, the cash flow statement would reflect this activity as a cash outflow from investing activities because it used cash. If the company decided to sell off some investments from an investment portfolio, the proceeds from the sales would show up as a cash inflow from investing activities because it provided cash.

Financing Activities

The third part of a cash flow statement shows the cash flow from all financing activities. Typical sources of cash flow include cash raised by selling stocks and bonds or borrowing from banks. Likewise, paying back a bank loan would show up as a use of cash flow.

---

[55] https://www.sec.gov/reportspubs/investor-publications/investorpubsbegfinstmtguidehtm.html

The following table is an example from ExxonMobil's 2016 cash flow statement:

	$MM
Cash flows from operating activities	
Net income including noncontrolling interests	8,375
Adjustments for noncash transactions	
Depreciation & Amortization	22,308
Deferred income tax charges/(credits)	(4,386)
Postretirement benefits expense in excess of/(less than) net payments	(329)
Other long-term obligation provisions in excess of/(less than) payments	(19)
Dividends received greater than/(less than) equity in current earnings of equity companies	(579)
Changes in operational working capital, excluding cash and debt	
Reduction/(increase) - Notes and accounts receivable	(2,090)
Reduction/(increase) - Inventories	(388)
Reduction/(increase) - Other current assets	171
Increase/(reduction) - Accounts and other payables	915
Net (gain) on asset sales	(1,682)
All other items - net	(214)
Net cash provided by operating activities	**22,082**
Cash flows from investing activities	
Additions to property, plant and equipment	(16,163)
Proceeds associated with sales of subsidiaries, property, plant and equipment, and sales and returns of investments	4,275
Decrease/(increase) in restricted cash and cash equivalents	-
Additional investments and advances	(1,417)
Collection of advances	902
Net cash used in investing activities	**(12,403)**
Cash flows from financing activities	
Additions to long-term debt	12,066
Reductions in long-term debt	-
Reductions in short-term debt	(314)
Additions/(reductions) in commercial paper, and debt with three months or less maturity	(7,459)
Cash dividends to ExxonMobil shareholders	(12,453)
Cash dividends to noncontrolling interests	(162)
Tax benefits related to stock-based awards	-
Common stock acquired	(977)
Common stock sold	6
Net cash used in financing activities	**(9,293)**
Effects of exchange rate changes on cash	(434)
Increase/(decrease) in cash and cash equivalents	(48)
Cash and cash equivalents at beginning of year	**$ 3,705**
Cash and cash equivalents at end of year	**$ 3,657**

Company Structure

The way a company is *legally structured* becomes an important criterion when evaluating investing in one company or another.

For U.S. Federal Income Tax purposes, there are two major types of corporate structures available:

- "C" Corp, which is a corporation that is taxed *separately* from its owners.

- Publicly Traded Partnerships, more commonly known as *Master Limited Partnerships* or MLPs, which are *pass-through* entities whereby income is taxed at the *individual* unit holder level, *instead* of being taxed at the corporate level.

What is a "C" Corporation?

A "C" Corporation is a commonly used corporate structure filing status in the United States. "C" Corporations have the following qualities[56]:

- Limited Liability protection, typically shareholders are not *personally* liable for the actions and debts incurred by a corporation.

- Separate entities, corporations are *separate* legal entities from the owners or shareholders.

- Taxation, corporations are taxed *separately* from its owners and incur a corporate income tax. For U.S. income tax purposes, shareholders' income from a corporation is only taxed whenever a corporation *pays out* a dividend to its shareholders. Corporations are in effect taxed *twice*, once income is generated inside the corporation, *then taxed again* once dividends are paid to shareholders. In the United States corporate level income taxes are currently 35% while dividends to shareholders can incur tax rates as high as 20%[57].

Most companies in the downstream sector are organized as regular "C" corporations, although there are certain refining Master Limited Partnerships (MLPs) in the market[58].

---

[56] https://www.bizfilings.com/toolkit/research-topics/incorporating-your-business/s-corporation-vs-c-corporation

[57] http://www.theasi.org/assets/EY_ASI_Dividend_and_Capital_Gains_International_Comparison_Report_2012-02-03.pdf

[58] CVR Refining is an example of a Refining MLP. Please visit
http://www.cvrenergy.com/CVRRefining/index.html

What is an MLP?

An MLP or Master Limited Partnership, in the energy industry, is a publicly traded limited partnership (PTP) that engages primarily in the energy sector, particularly in pipelines, fractionation, gas processing and other areas, particularly in the midstream space. An MLP, as a partnership, is a "pass-through" entity for U.S. Federal Income Tax purposes, meaning that taxes are not paid at the *entity* level, but instead are paid at the *unitholder* level. Each individual limited partner, as a unit holder, is an owner of the MLP, and thus owns a *pro-rata* share of the MLP. MLPs started to be more commonly used as a business entity after the 1986 reform of the Internal Revenue Code (IRC).

At the end of the year, each Limited Partner (LP) owner receives what is called a "K-1" schedule[59], which will show their pro-rata share of all *revenues*, *expenses* and *earnings* so that each unitholder can use this income information to file their personal U.S. income tax return. One of the key advantages of this tax structure is that "distributions" (distributions have roughly the same definition as dividends, but differ with regard to certain legal & tax ramifications) are *tax-free*, since taxes are already paid on the *accrual* accounting earnings. In other words, taxes are paid at the individual LP owner level on the MLP's *accrual accounting* earnings, not on *distributions* from the MLP to the owners. By being structured as an MLP, MLP entities avoid the recurrent issue with regular corporations that incur double taxation. In contrast, a regular "C corporation" pays income taxes at the *corporate level* on corporate income. Under a regular C-corporation, once dividends are distributed to shareholders, each individual shareholder is then taxed *again* on the dividend income received, which results in double taxation.

As can be seen previously, from a purely tax perspective, the MLP structure is significantly more tax-efficient since it allows the unit or shareholder to keep a higher percentage of the entity's earnings.

---

[59] http://www.tortoiseadvisors.com/mc/attachments/350/ABCs-of-MLPs_021816_Final.pdf

The following table provides an example of *how* an MLP structure is more tax-efficient than a regular C Corporation, assuming that 100% of earnings are distributed:

Item	MLP	C-Corp
Revenues	$100	$100
Cost of Goods Sold	($70)	($70)
Opex & SG&A	($10)	($10)
Depreciation & Amortization	($10)	($10)
Income before Tax	$20	$20
Corporate Income Tax	-0-	($7)
Earnings	$20	$13
Dividends	$20	$13
Individual Unitholder Partnership income Tax (assuming 35% personal rate)	$(7)	$0
Individual Shareholder Dividend Tax	$0	$2.60
After-Tax Cash flow to the Unit/Shareholder	$13	$10.40

Because of the depreciation expense allowance, 80% to 90% of the distribution a unitholder receives from a typical MLP is considered a *return of capital* by the Internal Revenue Service. A unitholder does not pay taxes immediately on this portion of the distribution. Instead, return of capital payments serve to reduce a unitholder's tax basis[60] in the MLP. This tax basis is used to determine gains and losses whenever the units are actually *sold.* Typically, an investor's initial basis is the price paid for the MLP units plus or minus certain adjustments (such as trading fees). Cash distributions *decrease* the tax basis, while the share of income from the MLP *increases* the tax basis or investment balance.

Because of this depreciation shield, 80% to 90% of the distribution a unitholder receives from the MLP is *tax-deferred*. The remaining piece of each distribution is taxed at normal income tax rates, not the special dividend tax rate. But the piece taxed at full income tax rates is typically only 10%-20% of the total cash distribution.[61]

---

[60] Tax Basis is generally the amount of a capital investment in property for tax purposes.
[61] https://www.investingdaily.com/13535/mlps-and-taxation-a-quick-refresher-for-tax-season/

Financial Metrics Overview

Financial metrics are any financial measures, which can be used to evaluate the financial performance of any business, across several industries and sectors in economy. These metrics are also used to analyze the overall market value, financial condition and performance of oil & gas companies. Although these metrics are based on the most commonly used definitions, many companies might define these metrics *significantly different* than what is used in this book and as such, where applicable, these differences are noted in the book.

The following financial metrics are used throughout this book:

- Market Capitalization
- Enterprise Value
- Total Consolidated Revenues
- Earnings or Net Income Attributable to the Corporation
- Gross Profit Margin and Percentage
- Net Margin
- Earnings Before Interest, Tax, Depreciation & Amortization or EBITDA
- Enterprise Value over EBITDA or EBITDA multiple
- Return on Capital Employed (ROCE)
- Cash Return on Capital Employed (CROCE)
- Return on Average Equity (ROE)
- Dividends or Distributions per Share or Unit
- Dividend Yield
- Capital Expenditures
- Cash flow from Operations (CFO)
- Free Cash Flow (FCF)
- Percent of Cash Flow from Operations devoted to Dividends/Distributions
- Total Shareholder Return (TSR)
- Debt to Equity Ratio
- Current Ratio
- Working Capital
- Price/Earnings Ratio
- Interest Coverage Ratio

Market Capitalization

Market capitalization, or market cap for short, is the notional value of a company's market value on a *specific date and time*. Market cap is calculated as the closing share price at a specific date, *multiplied by* the number of shares outstanding at the end of the same period. Market capitalization is based on market quoted prices, which fluctuate *every day*; therefore, market cap will *always* be fluctuating daily as well.

Market cap is a *notional* or *theoretical* value. If a company were to acquire another company in the market, the buying company would definitely not pay the current market cap since the share price would be *bid up* in *anticipation* of this acquisition.

Throughout this book the following methodology is used:

Ordinary shares *outstanding* as of December 31, 2016 (as per the company's SEC Form 10-K or form 20-F) *times* the closing price of the shares as of December 31, 2016[62]. Please note that some companies may provide shares outstanding at a later time period, such as January 31, 2017 or later, so that number of outstanding shares will be used *times* the closing price of such company's shares at December 31, 2016.

> *Corporation XYZ had 4,500,000 common shares outstanding as of Dec. 31, 2016 while the closing price for their stock was $40 per share, therefore the company's market cap as of Dec. 31, 2016 was 4.5MM shares times $40 which equals $180MM.*

Enterprise Value

Enterprise Value, or simply known as EV, is a commonly used metric to evaluate a company's total valuation assigned by both *equity holders* and *debt holders*. Enterprise Value is often used as an alternative to market capitalization, since it incorporates the value debt holders have assigned to a company *in addition* to the value assigned by equity holders.

As a side note, please note that the debt markets are typically far *larger* than equity markets[63], and have significantly more trading activity than equity markets. The trading activity of debt versus equity markets is usually a

---

[62] December 31st, 2016 fell on a Saturday, the closing price the day before is used

[63] As of the end of 2016, the outstanding value of all bonds in the United States was around $39 trillion, vs. an average valuation of US equity markets of about $26 trillion. Source: http://www.sifma.org/research/statistics.aspx

multiple of equity markets; therefore, the valuation placed by debt markets on a company needs to be factored in, which is what EV does.

Another reason why Enterprise Value is important is that it is considered as a *theoretical takeover price*[64]. In the event of a *buyout*[65], an acquirer would take on the company's debt, but would take cash from the acquired company, which is the logic behind why EV *subtracts* out cash and cash equivalents from the calculation. The final step is to *add back* minority or (non-controlling interest[66]). For simplicity purposes it is assumed that minority interest value is based on book value, since market value for a minority or noncontrolling interest may not be readily available.

Throughout this book Enterprise Value or EV is calculated as follows:

- Ordinary common shares *outstanding* at December 31, 2016 *multiplied by*
- Closing price of the common shares as of December 31, 2016, and the result, which is market cap *plus*
- Total value[67] of debt, both *current* and *non-current* as of the same period and as reflected in the company's balance sheet, including capital lease obligations *plus*
- Value of the *minority* interest or *non-controlling* interest *minus*
- Cash and cash equivalents

  Corporation XYZ, as of year-end 2016, had outstanding common shares of 100MM shares, a share price of $10, total debt and capital lease obligations of $200MM, minority interest market value of $100MM and cash and cash equivalents of $150MM. The company's enterprise value is calculated as follows, $1 billion market cap, plus $200MM, plus $100MM minus $150MM, resulting in an EV of $1,150MM.

Total Consolidated Revenues

Total Consolidated Revenues, also known as *gross revenues,* is a widely used metric to rank companies across *all industries* and understand their *relative* size. Fortune magazine uses gross revenues as a way to rank companies in

[64] https://valueandopportunity.com/2012/07/03/how-to-correctly-calculate-enterprise-value/
[65] A buyout is the purchase of a company's shares in which the acquiring party gains controlling interest of the targeted firm
[66] Minority interest is explained further on the next page
[67] The traditional formula for EV relies on the *market value* of debt, but for simplicity purposes we will use the *book value* of debt on the balance sheet

its Fortune 500 Company rankings[68]. Depending on the company, this metric might be called "gross revenues", "operating revenues" or "total consolidated revenues".

As a side note, overall large consolidated revenues do not necessarily translate into higher earnings. Traditionally, refining companies have much *higher* revenues than pure E&P companies, but *historically* refining companies have had significantly *lower* gross margins and thus *lower* earnings.

> *Corporation XYZ had total revenues of $4.5 billion for calendar year 2016, as per their reported income statement on their 10-K filings.*

Typically downstream companies include excise, sales and other fuel taxes as part of gross revenues and these are presented as a cost reduction later on the income statements[69]. These fuel taxes can be a significant percentage of total revenue, anywhere from 2% to 15% of total gross revenues recorded.

Earnings

Earnings or Net income attributable to the Corporation is a U.S. GAAP and IFRS[70] measure widely used to gauge the overall profitability of a corporation or entity. Net income reflects taxes, costs and expenses and is widely referred to in press releases as "earnings". This number *excludes* net income attributable to non-controlling interests (also called *minority interests*):

> *"A non-controlling interest, sometimes called a minority interest, is the portion of equity in a subsidiary not attributable, directly or indirectly, to a parent."[71]*

In other words, this is why this metric is called net income *attributable* to the entity or corporation, since it *excludes* the portion of earnings that the owners or shareholders of the corporation *do not have* a claim on.

---

[68] http://www.uspages.com/fortune500.htm

[69] Current U.S. GAAP guidance permits an entity to make an accounting policy election to present in-scope sales taxes on either a gross basis (included in revenues and costs) or a net basis (excluded from revenues). Source:
http://www.ifrs.org/Meetings/MeetingDocs/IASB/2015/March/AP07B%20Sales%20tax%20presenta tion%20Gross%20vs%20Net.pdf

[70] GAAP or Generally Accepted Accounting Principles are overseen by the Financial Standards Board or FSB and are required for US-based companies issuing financial statements. IFRS or International Financial Reporting Standards are overseen by an International Board and many countries have adopted these standards. The SEC has allowed foreign issuers to use IFRS instead of having to convert to U.S. GAAP. At its public meeting in Washington on 15 November 2007, the SEC voted to allow foreign companies to submit financial statements to the Commission using IFRSs as adopted by the IASB without having to include a reconciliation of the IFRS data to US GAAP. https://www.iasplus.com/en-gb/resources/regional/sec

[71] http://www.fasb.org/summary/stsum160.shtml

Corporation XYZ had total company wide net income of $100MM, while net income attributable to non-controlling interests was $20MM, therefore, net income attributable to XYZ or earnings were $80MM.

In the case of U.S. Master Limited Partnerships it is important to understand that their quoted "Earnings" do not include the effects of U.S. Federal corporate income tax.

Adjusted Earnings

Earnings may have to be adjusted to account for "Special items", which are items such as asset gains or asset impairments that do not reflect "*core earnings*" but are still accounted under GAAP rules as earnings. These items tend to be called special items since they are typically non-recurring and therefore do not reflect the true earning power of the company being evaluated. These special items are included in GAAP earnings but have to be adjusted to that earnings reflect *on-going* core operational earnings:

- Asset impairments
- Accruals for pending claims or settlements
- Certain Tax impacts
- Gains on sale of assets
- Restructuring
- Termination of contracts

Corporation XYZ had earnings of $200MM in 2014, $220MM in 2015 and $500MM in 2016. Included in earnings in 2016 was a $240MM gain related to a sale of a pipeline, and since gains are not part of core earnings need to be excluded. Therefore adjusted earnings for XYZ in 2016 were $500MM minus $240MM equal to $260MM.

It is important to note that a company's 10-K fillings need to expressed in terms of *GAAP earnings* and cannot include any reference to special items. Adjusted earnings are usually quoted on the company's press release, earnings release presentation and other materials, but cannot be in the quarterly or yearly fillings with the SEC[72].

---

[72] The regulatory requirements for filings with the SEC are quite extensive, for more information visit this website: http://www.pwc.com/us/en/cfodirect/assets/pdf/in-brief/us-2016-22-sec-non-gaap-interpretive-guidance-update.pdf

Gross Margin & Gross Operating Margin

Gross margin is a very important metric in the Oil & Gas industry, particularly in the midstream and downstream sectors. Since the downstream business is largely a *margin* business, it is important to understand how gross margin is calculated and arrived at.

Throughout this book gross margin is calculated as follows:

- Revenues (including equity earnings) *minus* purchases or cost of goods sold *equals* Gross Margin.

Gross operating margin is calculated as follows:

- Gross Margin is then further reduced by *subtracting* operating, general and administrative expenses to arrive at a gross operating margin.

  XYZ Company had 2016 revenues of $1,500MM, equity earnings of $100MM, purchases of $1,200MM. Therefore, gross margin is equal to $1500MM plus $100MM minus $1200MM, which equals to $400MM.

Many companies then *subtract* out operating and SG&A expenses from *gross margin* since several types of assets do not have substantial "cost of goods sold" or do not involve the *buying* and *selling* of commodities, but instead provide a service, such as transportation, terminaling and storage of hydrocarbons. Pipelines are a good example of this case, whereby a pipeline would have very little "cost of goods sold" but more *operating* costs. This is what is commonly called *gross operating margin*.

To calculate gross *operating* margin:

  In 2016, XYZ Company had gross margin of $400MM and operating expenses of $100MM while G&A expenses were $50MM. Therefore, gross operating margin in 2016 was $250MM.

Gross Margin Percentage

As previously mentioned, the downstream business is a *margin* business, unlike oil & gas exploration & production. Gross margin percentage in this book is calculated as gross margin *divided* by revenues to arrive at a gross margin percentage.

  Corporation XYZ had revenues of $100MM, cost of goods sold of $95MM, thus its gross margin was $5MM, indicating a gross margin percentage of 5%.

Net Margin

Net margin is calculated by taking net income or earnings and dividing it by the revenues to arrive at a percentage. Historically, the downstream business tends to have very low profit margins as compared to cells. The net margin can also be expressed in terms of cents per dollar, in other words for $1 of sales that a company their profit margin can be 2 cents.

Corporation XYZ had revenues of $5 billion, while net income or earnings were $100 million. $100MM divided by $5000MM equals to a 2% net margin. In other words, for every $1 of sales, XYZ kept 2 cents as net margin.

EBITDA

EBITDA or Earnings Before Interest, Taxes, Depreciation & Amortization, is a commonly utilized *Non-GAAP*[73] metric, particularly in the midstream sector of the oil & gas industry. There are several reasons why EBITDA is widely used in evaluating a company's earning potential:

- To a large extent, EBITDA reflects the underlying company's operating income and performance, by taking out of earnings *non-operating* items such as taxes, depreciation and interest.
- Can be used as a proxy for operating cash flows, since depreciation is usually the largest *non-cash item* in a company's income statement.
- For tax "pass-through" entities like Master Limited Partnerships that do not have corporate income taxes, EBITDA is a more comparable metric to the cash generation potential of the entity
- Income taxes can fluctuate significantly for non-operating reasons, such as tax credits, deferred taxes as well as legislation.

For the purposes of this book EBITDA is calculated as follows:

- Start with GAAP net income attributable to the entity *plus*
- Minority interest or non-controlling interest *plus*
- Interest expense *plus*
- Income tax expense *plus*
- Depreciation & Amortization
- Equals EBITDA

   Corporation XYZ had net income attributable to XYZ of $600MM, while non-controlling interest was $100MM, thus XYZ had total net income of

---

[73] https://www.sec.gov/divisions/corpfin/guidance/nongaapinterp.htm

$700MM. Interest expense was $50MM, income tax was $150MM while depreciation & amortization was $200MM, thus EBITDA was $1,100MM for that fiscal year.

EV/EBITDA or EBITDA Multiple

Enterprise Value over EBITDA, also known as EBITDA multiple, is a valuation metric very commonly used to assign a market valuation to *individual assets, operating segments* or even *entire companies.* EBITDA multiples become very critical when assessing the valuation of a company with multiple businesses or reporting segments. For example, for an integrated oil & gas company with businesses in the upstream, midstream & downstream, the EBITDA multiple is necessarily different for each of its reporting segments depending on investors' preferences, therefore its composite EBITDA multiple will likely be much different than a pure play company in the refining or E&P space.

This metric is calculated as follows:

- Enterprise value of a company *divided by*
- Earnings Before Interest, Taxes, Depreciation and Amortization

  In 2016, Refining Company XYZ had EBITDA of $200MM while its Enterprise Value or EV was $2 billion at year-end. Taking $2 billion and dividing it by $200MM gets an EV multiple of 10.

One key advantage of using EBITDA multiples is that it allows a much *faster* analysis of an asset or company's value by using the concept of *comparable equities.* For example say that an analyst is trying to assess the value of Refining Company A vs. Refining Company B. The traditional method would be to estimate all the future *revenues, expenses* and *capital expenditures* of the company, then use a discounted cash flow (DCF) model to arrive at a valuation in today's dollars or present value for each company. A simpler method would be calculate the currently traded EBITDA multiples (based on market values) for similar companies and use this EBITDA multiple with Refining Company A's EBITDA to obtain a quick and high level estimate of how much this company could potentially be worth in the stock market. The same concept applies to valuing individual assets against another.

Refining Company ABC, which is not publicly traded, has EBITDA of $400MM and Total Debt of $1billion. By using a similar multiple of Refining Company XYZ, which is 10, we arrive at a hypothetical Enterprise

Value of $4 billion, implying that a possible market capitalization of $3 billion if ABC were to pursue an Initial Public Offering after deducting $1 billion of debt from Enterprise Value.

Return on Capital Employed (ROCE)

Return on Capital Employed or ROCE is a widely used financial ratio that measures how *efficiently* a company can generate profits with their current capital employed or net assets. Businesses with consistently high ROCE tend to be more highly valued over long periods of time than businesses with low ROCE. ROCE has 2 main components, ROCE income and average capital employed.

Throughout this book, ROCE is calculated as follows:

- ROCE income: Net Income Attributable to the Corporation *plus* Non-Cash Adjustments *plus* after-tax interest expense *plus* minority or non-controlling interest.
- Average capital employed is calculated by taking the average of beginning *and* ending Total Assets *minus* the average of beginning *and* ending Current Liabilities.
- ROCE income is then *divided* by the average capital employed to arrive at ROCE.

  Corporation XYZ had net income of $80MM and after-tax interest expense of $20MM in 2016. Therefore ROCE Income is equal to $100MM. Corporation XYZ had total beginning and ending total assets of $1,400MM and $1,500MM respectively. XYZ had beginning and ending current liabilities of $500MM and $600MM respectively. Therefore, average capital employed for 2016 was $900MM. ROCE income of $100MM is divided by average capital employed of $900MM to arrive at a Return on Capital Employed or ROCE of 11.1%.

For companies with a large amount of capital employed, i.e. midstream companies, using ROCE as a metric may not be the most optimal measure of economic evaluation.

A pipeline is a good example of an asset having a *large* capital employed but having a *low return* on capital employed. One area that ROCE does not factor in is in the *stability* of these cash flows. ROCE for refining assets can fluctuate *widely* from one year to the other since margins can change drastically from one year to the other. Another limitation with ROCE is that ROCE can be lower whenever a company embarks on long-term projects, spanning for years, for example an LPG terminal or building a new refinery since these assets increase Property, Plant & Equipment but are basically *"pre-productive capital"* since these assets are not yet operational and thus do not generate earnings *yet*. Companies with no recent capital expenditures and long depreciated assets will tend to have higher ROCE, which may indicate future slower growth. The following table provides a sequence of why ROCE would decline due to the *pre-productive capital* effect:

Year	Capital Employed	ROCE Income	ROCE	Comment
1	$1,000	$100	10%	Company announces investment for a new asset
2	$1,500	$120	8%	Company invests in long-term project which increases capital employed
3	$2,500	$150	6%	Investments in long-term project continues, increasing capital employed and no earnings have been generated yet from the new project
4	$3,000	$300	10%	New project becomes operational mid-year, ROCE income increases due to projects
5	$2,900	$450	16%	New project continues to generate income and capital employed starts going down as assets depreciate

Cash Return on Capital Employed (CROCE)

Similar to Return on Capital Employed, Cash Return on Capital Employed or CROCE is a financial ratio that measures how *effective* the company is in generating *cash* from its given average capital employed. Similar to ROCE, CROCE has 2 main components, CROCE numerator and average capital employed. The only difference in the calculation is that for CROCE the numerator or ROCE income adds back Depreciation and Amortization or D&A to arrive at a comparable *cash* numerator instead of an *income* numerator.

The reason D&A is added back is that depreciation tends to be the largest *non-cash* item in accrual accounting and provides a number that is closer to cash earnings. Companies cannot use cash flow from operations as the numerator since it includes fluctuations related to working capital movements.

Throughout this book, CROCE is calculated as follows:

- CROCE numerator: Net income attributable to the Corporation *plus* Non-Cash Adjustments *plus* After-tax Interest expense *plus* Minority Interest *plus* D&A.

- Average capital employed is calculated by taking the average of beginning *and* ending Total Assets *minus* the average of beginning *and* ending Current Liabilities.
- CROCE numerator is then *divided* by the average capital employed to arrive at CROCE.

Corporation XYZ had net income of $80MM, after-tax interest expense of $20MM and D&A of $30MM in 2016. Therefore CROCE numerator is equal to $130MM. Corporation XYZ had total beginning and ending total assets of $1,400MM and $1,500MM respectively. XYZ had beginning and ending current liabilities of $500MM and $600MM respectively. Therefore, average capital employed for 2016 was $900MM. CROCE numerator of $130MM is divided by average capital employed of $900MM to arrive at a Cash Return on Capital Employed or CROCE of 14.4%.

Return on Average Equity (ROE)

Return on Average Equity or ROE is another measure of performance and shows how *profitable* or *efficient* a company is in generating returns using shareholders' capital or equity. ROE measures how effective the company is in generating returns given its *current* level of total equity. In other words, this metric tells investors how profitable the company is given the invested capital *actually owned* by shareholders, common or minority owners, instead of debt holders. ROE has one advantage over ROCE in the fact that it is easily defined and quite comparable against other different industries. ROCE on the other hand, has one comparability issue primarily due to the definition of average capital employed, which may vary substantially from one company to the other. Throughout this book, *average* total equity is used, which includes non-controlling interest's equity. Average total equity is calculated by adding the beginning and ending balances of total equity and then dividing by two to arrive at an *average* equity. ROE has two components, the first ROE numerator and the second one Total Equity.

Throughout this book, ROE is calculated as follows:

- ROE numerator: Earnings or Net income attributable to the company *plus* net income attributable to non-controlling interests.
- Total average equity: Beginning total equity *plus* ending total equity and the product of this is *divided* by two to arrive at an average for the period.
- ROE numerator is then *divided* by total average equity to arrive at ROE.

In 2016 Corporation XYZ had net income attributable to XYZ of $150MM, while net income attributable to non-controlling interests was $50MM, therefore total net income or ROE numerator was $200MM. Beginning total equity was $1,000MM, while ending total equity was $1,100MM. Total average equity is thus $1000MM plus $1100MM and the sum of this divided by two, which equals to $1050MM. ROE numerator of $200MM is then divided by total average equity of $1050MM to arrive at a return on equity or ROE of 19%.

Dividends or Distributions per Share/Unit

Dividends are distributions of property a corporation pays to the shareholders, whether in cash or in stock. Most companies listed on major stock exchanges declare and pay dividends on a quarterly basis. Several companies pay dividends on a yearly basis and are subject to approval by their board of directors. Dividends are usually stated in terms of a period of time:

Corporation XYZ's Board of Directors declared a quarterly dividend of $0.25 per share for the Second Quarter of 2017. The dividend will be paid to shareholders of record on August 1, 2017.

Many foreign oil & gas companies, especially some of the non-U.S. and non-Western European ones, *do not offer* an ordinary dividend per quarter. Instead, these companies' dividends are declared and approved by their respective board of directors once-a-year. These dividends might fluctuate widely from year to year based on several economic factors as well as management decisions. In contrast, the more established companies offer attractive and constant quarterly/yearly dividend payments that are more or less *fixed* and highly *predictable*.

As an example, Petrobras' board of directors approves and declares a yearly dividend payment, which is based on several factors (cash-flows from operations, upcoming debt payments, capital expenditures and other criteria) and then proceeds to pay this dividend once-a-year. Petrobras' board of directors met in April 2014 and approved an annual dividend of $0.4799 for common ADR shares[74]. Since the last payout, Petrobras has not paid a dividend since 2014, due to the low price of oil as well as internal management issues at Petrobras[75].

---

[74] http://investidorpetrobras.com.br/en/notices-and-facts/material-fact-payment-of-interest-on-own-capital-13.htm
[75] http://www.investidorpetrobras.com.br/en/shares-and-dividends/dividends

In contrast, ExxonMobil's dividends per common share were $0.40 *per quarter* in the 4th quarter of 2008 and were subsequently increased *every year*, with the most recent increase being in the 2nd quarter of 2017 to $0.77 per share[76]. ExxonMobil's dividends per quarter are expected to continue to be $0.77 per share until next year when the board of directors will likely increase it. The possibility of the dividend being *decreased* or *not declared* is highly unlikely for an established company like ExxonMobil, and in fact, ExxonMobil has increased its dividend payment by 6.4% per year for the last 34 years[77].

Dividend Yield

Dividend yield is calculated on an annual basis and it is simply the dividends per year *divided* by the current share price of the company. The share price used throughout this book is the closing price as of December 31, 2016. Dividends paid in a year are annualized, in other words if a company paid $0.25 per share every quarter, that translates to $1.00 per share per year.

> *Corporation XYZ had yearly dividends of $1.00 per share in 2016, while its December 31, 2016 year-end share price was $10.00 per share. Therefore XYZ's dividend yield is simply $1.00 per share divided by $10.00 per share or 10%.*

Dividend yield is one of the most widely used metrics in the financial industry to evaluate the income or *yield* ability of a company. In the current stock market environment, many companies' dividend yields are favorable in comparison to historical averages and other investments such as bonds.

Traditionally, oil & gas companies or investments can be ranked in terms of dividend yields, from *lowest* yields to *highest* yields as follows:

- Independent E&P companies (usually *zero* dividends)
- Service companies
- Independent refining companies
- U.S. Integrated oil & gas companies
- Midstream companies
- Royalty Trusts
- European integrated oil & gas companies that due to their dividend policies, European companies have historically distributed more to

[76] http://news.exxonmobil.com/press-release/exxon-mobil-corporation-declares-second-quarter-dividend-7

[77] http://corporate.exxonmobil.com/en/investors/stock-information/dividend-information/overview

shareholders in the form of dividends and less as share repurchases than their U.S. peers

Capital Expenditures

In a capital intensive business like the oil & gas industry, capital expenditures are critical in order to continue to provide future earnings to shareholders. CAPEX is particularly important in the E&P sector, whereby CAPEX is a necessity in order to manage and reduce natural field *declines* in oil & gas production. In the Downstream, CAPEX is also critical, but *large amounts* of CAPEX are not necessary to the extent that an E&P requires. One of the primary drivers of CAPEX in Refining is driven by *regulatory compliance* of ever *more stringent* fuel specifications, particularly in sulfur and emission controls, whereby an existing refinery requires additional capital expenditures to upgrade equipment and units and comply with these new fuel requirements, such as Tier 3 fuels[78] or ultra-low sulfur fuels[79].

Capital expenditures figures are found on the investing section of the cash flow statement, similar to the example below:

Cash Flows From Investing Activities	$MM
Capital Expenditures and investments	(1000)
Proceeds from asset dispositions	500
Advances/loans – related parties	(100)
Collection of advances/loans related parties	60
Other	30
Net cash provided by (used in) investing activities	**(510)**

Traditionally, the following sectors of the oil & gas industry have been the most capital intensive, requiring a continuous high level of capex to *sustain* and possibly *grow* earnings:

- Exploration & Production sector requires large amounts of capex over prolonged periods of time in order to reduce natural field decline and grow production over time. This sector of the industry has traditionally required ongoing CAPEX which increases with the overall level of activity in the industry. This sustained level of CAPEX over long periods of time can impact an E&P company's Free Cash flow (discussed further in the chapter).

---

[78] http://analysis.petchem-update.com/operations-maintenance/us-refineries-focus-epa-compliance-delay-other-spending-due-lower-margins
[79] http://www.hydrocarbonprocessing.com/magazine/2016/february-2016/trends-and-resources/business-trends-clean-fuels-a-global-shift-to-a-low-sulfur-world

- Midstream sector (including pipelines, LNG plants), requires significant capex at the *initial* phase whenever a pipeline, fractionator, gas plant or other midstream asset is built. Over time, capex is also needed to grow earnings by adding or developing more assets. Once a pipeline, LNG plant or any other midstream asset is built, the capex required to maintain the asset in operating condition is *relatively* small compared to the initial investment. An initial investment in a midstream asset can be in the billions or even tens of billions of dollars with these assets having very long term useful lives (usually in the 3-5 decade range or even more).

- Service companies sector, depending on which sector a service company serves may have low to medium levels of capital expenditures. Those service companies that lease out equipment, such as drilling equipment, will usually have higher capex than those servicing downstream companies.

- Downstream, similar in a way to Midstream, has a relatively modest *on-going* CAPEX. Once a refinery or terminal is built, which requires a large initial investment, the CAPEX required to maintain the asset in operational condition is *relatively modest* compared to the initial investment. Certain short-term projects can be done to debottleneck existing refinery's units to allow increased processing of certain types of crudes with modest to medium capital requirements. In the U.S. recent examples can be seen of these expansions, where several refineries are adding or expanding their crude distillation units (CDU) to allow for increased processing of growing lighter crude oil production in the U.S. Similar to LNG plants, once a refinery is built, extended maintenance or what are called "turnarounds" expenditures are *relatively* small when compared to the initial investment to build the refinery. For downstream, in general terms, it can be said that their *on-going* capex is relatively modest when compared to E&P operations.

Cash Flow from Operations (CFO)

Cash Flow from Operations or CFO is simply the cash generated from the company's operations. CFO is an indicator of the company's ability to fund capital expenditures as well as fund shareholder distributions.

Cash Flow from Operations is a widely used metric throughout the world of finance. Operating Cash Flows are reported on a company's cash flow statement, on the operating section.

CFO is *generally* calculated as follows:

- GAAP Net Income *plus*
- Depreciation & Amortization *plus*
- Net foreign currency effects
- +/- Deferred Income adjustments
- +/- Changes in working capital
- +/- Net Gain or Loss on asset sales

Any company that provides a Statement of Cash Flows would already provide cash flow from operations, so there is no need to calculate this metric.

Free Cash Flow (FCF)

Free cash flow is a *non-GAAP* financial metric used to analyze how much cash flow is available for distribution to shareholders or debt holders. Free Cash Flow or FCF is also used to gauge the overall cash that is being generated from the company's operations.

Free Cash Flow is calculated as follows:

- Cash from Operations (from the Cash Flow statement in the Operating Section) *minus* capital expenditures

  Corporation XYZ had Cash Flow from Operations (CFO) in 2016 of $200MM and capital expenditures of $50MM; therefore Free Cash Flow is $150MM.

Some companies further refine this FCF metric by taking out only *"sustaining capital expenditures"*, which are those that keep the *existing* equipment and assets in operating condition but do not add further improvements or increase in throughput.

Percent of Cash Flow from Operations devoted to Dividends or Distributions

This metric allows an investor to quickly gauge how *safe* or *predictable* the dividend or distribution will be. Companies with a low percent of CFO devoted to dividends have more financial room to keep the dividend at the same level, including during challenging economic environments, with *high quality* companies having the ability to increase their dividend payments over time. For the purposes of this book, Percent of Cash Flow from operations devoted to dividends is calculated as follows:

- Total Dividends or Distributions *divided* by Cash Flow from Operations

  Corporation XYZ, as per their cash flow statement, paid dividends of $100MM to shareholders in 2016, while cash flow from operations was $500MM. Therefore, the percent of cash flow from operations devoted to dividends in 2016 was 20%.

Before oil prices started to decrease substantially in mid-2014, oil & gas companies, particularly E&P companies, devoted a low percentage of cash flow from operations to dividends. In the E&P sector, these cash flows are required to maintain the base business and expand production; therefore a relatively low percentage of cash flows from operations would be devoted to dividends or other shareholder distributions.

In the downstream business, particularly in the United States, refining & marketing companies have been able to not only maintain but actually increase their dividend payments over the past couple of years. Other companies with a significant downstream exposure, such as ExxonMobil, have been able to increase their dividend payments during these challenging times for the oil & gas industry. Such is the case with an Independent Downstream company like Marathon Petroleum or MPC. Since Marathon Petroleum or MPC was spun off from Marathon Oil, MPC has been able to increase their dividend from $0.10 in 2011 to $0.36 in 2017, an impressive increase of more than three times[80]. Even more impressive has been Valero, the world's largest independent refining company, which has increased its dividend per share from $0.05 in 2011 to $0.70 per share in 2017, an increase of more than 1000%[81].

Total Shareholder Return (TSR)

Total Shareholder Return or TSR is one of the most widely, if not the most widely, used financial metric to evaluate companies across many different sectors. Total Shareholder Return measures the *increase* in wealth shareholders have experience over a period of time from owning shares in a company.

---

[80] http://ir.marathonpetroleum.com/phoenix.zhtml?c=246631&p=irol-dividends
[81] http://www.investorvalero.com/phoenix.zhtml?c=254367&p=irol-dividends

Total Shareholder Return or TSR throughout this book is calculated as follows:

- Share price at year-end *minus* share price at the beginning of the year *plus* dividends paid during the year.
- The result above is then *divided* by the share price at the beginning of the year.

    At December 31, 2016 Corporation XYZ's share price was $50.00 while the share price at December 31, 2015 was $40.00. XYZ paid dividends to shareholders of $4.00 per share during 2016. Therefore $50 minus $40 plus $4.00 equals $14, which is then divided by $40 per share to arrive at a TSR of 35% in 2016.

As of mid-year 2017, downstream and integrated oil & gas companies are offering relatively high dividend yields[82]:

Company	Type	Dividend Yield
Royal Dutch Shell	IOC	6.91%
BP	IOC	6.54%
Total	IOC	5.24%
Valero	Downstream	4.48%
Chevron	IOC	4.13%
Eni	IOC	3.89%
ExxonMobil	IOC	3.78%
Phillips 66	Downstream	3.62%
Marathon Petroleum	Downstream	2.73%

The largest component of TSR in any company usually comes from dividend income growth *compounded* over several years.

Please note that the TSR calculated in this book *does not* assume that dividends are reinvested for the 1-yr period. Historically, a large portion of the total shareholder return of companies has been achieved in any sector through the *reinvestment* of dividends into purchasing *more* shares instead of "cashing out dividends" as income.

Debt to Equity Ratio

Debt to Equity ratio is a balance sheet metric that serves to convey how much *leverage* or *borrowed* funds a company is using *versus* owners' equity to grow the business. A high debt-to-equity ratio over a *sustained period* of time

---

[82] Source: www.dividend.com as of May 27, 2017

would indicate that a company's balance sheet would be *less* strong than its competitors. The metric in this book incorporates both the portion of long-term debt that is due within one year plus total long term debt. The equity portion of the metric is total equity, which includes non-controlling interests' portion of equity. To calculate debt-to-equity ratio:

- Total debt, including *both* the current portion of long-term debt *plus* long-term debt *divided* by total equity.

   Corporation XYZ in 2016, as per their balance sheet, had current portion of long-term debt of $100MM while long-term debt on the long-term liabilities section of the balance sheet was $900MM. XYZ had year-end equity balance of $5,000MM, which includes a non-controlling interest portion of $500MM. Therefore, XYZ debt-to-equity ratio is $1,000MM divided by $5,000MM which results in a debt-to-equity ratio of 20% at year-end 2016.

Current Ratio

The current ratio is a balance sheet ratio that measures a company's ability to meet *current short term* obligations (considered to be paid within one year), such as meeting payroll for employees, paying vendors or paying other short-term obligations due within one year. The current ratio measures how much *liquidity* or cash a company has.

The current ratio is calculated as follows:

- Current Assets *divided by*
- Current Liabilities

   Corporation XYZ had total current assets of $1 billion while current liabilities were $800MM at year-end 2016. Taking $1 billion and dividing it by $800MM arrives at a ratio of 1.25

Typically, having a current ratio lower than 1.00 indicates a probability of the company having problem meeting its short term obligations, although the company could use refinancing mechanisms, such as issuing long term debt to pay for current debts due to improve this ratio. On the contrary, having a very high current ratio is not necessarily a good indicator either, since it probably means that the company is not managing its working capital very well or using cash efficiently[83].

Working Capital

Similar to the current ratio, working capital is a liquidity metric that provides an insight to an investor as to whether a company has sufficient liquidity to operate in the short term. In addition, working capital provides an insight into how *efficiently* a company is partaking in the cash conversion cycle, which measures how well a company collects cash from customers and pays outs cash to vendors and other creditors.

Working capital is calculated as follows:

- Current Assets, which include accounts receivable and inventories *minus*
- Current Liabilities, which includes accounts payable

One of the key factors impacting working capital in the downstream business is payment terms for different commodities. Typically, in the U.S., crude purchases have payment terms of around 30 days, while fuel sales by refiners are collected from customers within less than 1 week, creating a positive working capital impact for downstream companies. This is in contrast to the upstream business, which usually has similar payment terms for its sales (crude sales at 30 day payment terms) than for its cash outflows, primarily capex vendor and royalties, which are also 30 days, creating a flat to slightly negative working capital impact.

Another key point impacting downstream, and in particularly, the Refining business, is commodity prices. In a rising crude oil price environment, refiners would usually tie out every month more working capital for crude purchases and inventory than the prior month, and if it's not followed by a rising fuel price environment, would result in a negative working capital impact. The subject of working capital management has been increasingly in the spotlight as volatile commodity prices impact collections and creditworthiness of many companies in the oil & gas industry, with a particular impact to the oilfield services companies[84].

---

[83] https://www.boundless.com/accounting/textbooks/boundless-accounting-textbook/reporting-of-current-and-contingent-liabilities-9/reporting-and-analyzing-current-liabilities-64/current-ratio-302-3754/

[84] http://www.ey.com/gl/en/industries/oil---gas/ey-cash-in-the-barrel-2014-working-capital-performance-by-segment

Price Earnings Ratio

The price earnings or P/E ratio as is commonly known is a very simple but *powerful ratio* to use when evaluating companies. The P/E ratio can indicate how relatively *undervalued* or *overvalued* a company's share price is.

The price earnings ratio is calculated by taking a company's share price and *dividing it by* its Earnings per Share or EPS in a year. For the purposes of this book, EPS used is for fiscal year 2016 while the share price is as of December 31, 2016.

> *Corporation XYZ had 2016 Earnings-per-Share or EPS of $2 while its share price at year-end was $ 10, therefore XYZ's P/E ratio was 5.*

Interest Coverage Ratio

The interest coverage ratio provides investors and creditors a gauge of how effectively a company can cover the interest payments on its current debt as well as be able to take on additional debt.

This ratio is calculated by taking Earnings Before Interest & Taxes or EBIT and *dividing* it by total interest expense. Please note that *interest expense* should also include *capitalized interest*. Capitalized interest is the cost of the funds used to finance the construction of a long-term asset that an entity constructs for itself. The capitalization of interest is required under the accrual basis of accounting, and results in an increase in the total amount of fixed assets presented on the balance sheet[85]. Therefore, to get an accurate picture of the actual *cash paid* for interest, capitalized interest, which is found on the cash flow statement, must be *added back* to interest expense, found on the income statement.

> *Corporation XYZ had Earnings Before Interest & Taxes of $100MM, while its total interest expense for the year was $19MM. The company also incurred in the same year, $1MM of capitalized interest, therefore total "cash" interest was $20MM. EBIT of $100MM is then divided by $20MM, resulting in a very healthy interest coverage ratio of 5.0.*

---

[85] https://www.accountingtools.com/articles/what-is-capitalized-interest.html

Chapter III – Petroleum Refining

"In times of rapid change, experience could be your worst enemy."– J. Paul Getty

Refining

Refineries convert raw crude oil and other feedstocks into *refined petroleum* products of *higher* value. These refined products are used throughout the world in a variety of industries, such as in the transportation sector, consumer goods, electronics, and many other products. Below is a typical breakdown of products that can be made from one barrel or 42 U.S. gallons of crude oil[86]:

Typical Product	Gallons	Percent of total
Gasoline	19.74	47%
Diesel Fuel & Heating Oil	9.66	23%
Other Products	7.56	18%
Jet Fuel	4.2	10%
Liquefied Petroleum Gas or LPGs	1.68	4%
Asphalt	1.26	3%
Total	**44.1**	**105%**

As can be seen from the table above, the volume of products being produced out of the refineries is *higher* than the 42 gallons of crude oil being processed. This is due to what are called *volumetric gains*, which are covered further in the chapter[87].

Refining crude oil is a *margin* business since it depends primarily on the difference in price between *inputs* (Crude oil and feedstocks) and *outputs* (refined products, primarily motor gasoline, diesel and jet fuel). The following chart provides prices per barrel for crude oil, gasoline and heating oil or diesel since 1986. The bigger the *gap* or *spread* between the price of crude oil and refined products, the higher the profitability of refining:

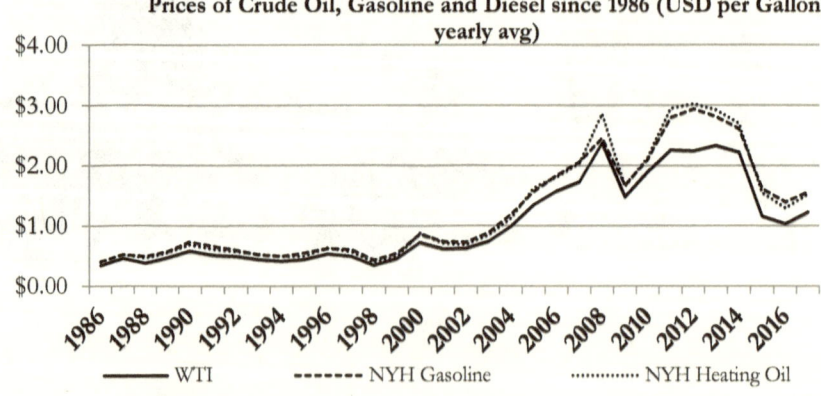

Prices of Crude Oil, Gasoline and Diesel since 1986 (USD per Gallon, yearly avg)

——— WTI   ------- NYH Gasoline   .... NYH Heating Oil

---

[86] http://www.eia.gov/energyexplained/index.cfm?page=oil_home
[87] Volume gains, as explained further in this chapter, are related to the different densities of products vs. crude oil.

There have been several times in history when crude prices continue to *increase* yet refined products prices stay *flat* or even *decline*. Such was the case in the years of the global economic crisis of 2008-2010, when refined product prices *declined* substantially due to an oversupply of refined products with too many refineries in operation. After this crisis, from 2011-2014, several refineries were shut down, which helped improve refining margins during this period, especially in the U.S. The refining business continues to be a very competitive environment and will remain a cyclical business for the foreseeable future[88]:

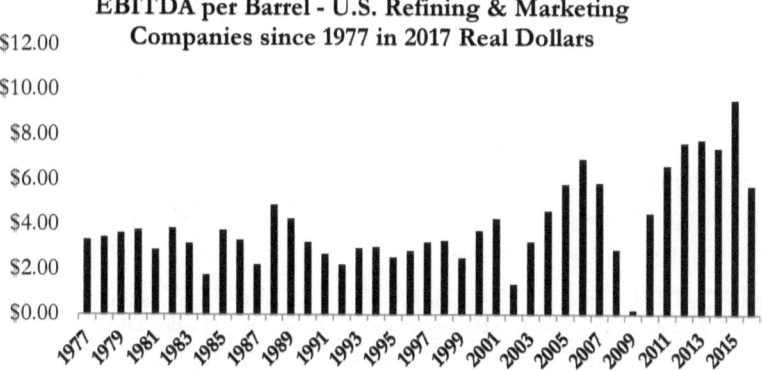

One positive factor that has occurred over the past couple of years, particularly for U.S. refineries, has been the increase of crude oil production in the U.S. with no corresponding outlet for these crude oils. As refined products demand in the U.S. has been *flat* or *declining* in some areas, U.S. exports of refined petroleum products have experienced a boom, going from about 1.3MMBPD in 2007 to close to 3.4MMBPD in early 2017.[89]

Thanks to this rising U.S. crude oil production, U.S. refiners, compared to their international counterparts, have experienced over the recent years what is called a *feedstock advantage*. What does this mean? A feedstock advantage is, being able to produce the same commodity product, but from *less expensive* inputs or feedstock sources. For example, say that the *same* or *similar* type crude oil with the *same* quality can be purchased in the US at a *discounted* price due to lack of takeaway capacity in a particular basin, this would provide domestic refiners that purchase this discounted crude oil an advantage over international refiners.

---

[88] Source: EIA Financial Reporting, plus select Refining companies since 2011 (VLO & MPC)
[89] https://www.eia.gov/dnav/pet/hist/LeafHandler.ashx?n=pet&s=mtpexus2&f=m

This feedstock advantage in the U.S. over the past couple of years has been caused by record production of *both* U.S. crude oil and natural gas, which has given US refineries an inherent *competitive advantage*. With U.S. refineries having significant feedstock advantage and *lower energy costs*[90] than their international counterparts, future competitiveness of U.S. refineries looks very promising.

Definitions

Before we being our journey of learning about Refining processes, the business itself and metrics, it might be valuable to provide a couple of definitions and concepts. We'll start by defining the different types of hydrocarbons, then define what constitutes gasoline, diesel and jet fuel, and then discuss specific properties that these three products have to meet.

Types of hydrocarbons

Let's review the different types of hydrocarbons that exist. The naming convention of hydrocarbons depends on the number of carbon atoms, chemical structure and the type of bond:

Number of Carbon Atoms	Prefix Convention	Single Bond Hydrocarbons	Chemical Structure	Double Bond Hydrocarbons	Chemical Structure
1	*Meth*	Methane	CH_4	N/A	N/A
2	*Eth*	Ethane	C_2H_6	Ethylene	$H_2C=CH_2$ or C_2H_4
3	*Prop*	Propane	C_3H_8	Propylene	$H_2C=CHCH_3$ or C_3H_6
4	*But*	Butane	C_4H_{10}	Butylene	$CH_3CH_2CH=CH_2$ or C_4H_8
5	*Pent*	Pentane	C_5H_{12}	Pentene	$H_2C=CHCH_2C$ H_2CH_3 or C_5H_{10}
6	*Hex*	Hexane	C_6H_{14}	Hexene	$CH_2=CH(CH_2)_3$ CH_3 or C_6H_{12}
7	*Hept*	Heptane	C_7H_{16}	Heptene	$CH_2=CH(CH_2)_4$ CH_3 Or C_7H_{14}
8	*Oct*	Octane	C_8H_{18}	Octene	$CH_3(CH_2)_5CH=$ CH_2 Or C_8H_{16}

---

[90] Lower energy costs are a significant source of advantage for a refinery. Energy costs are discussed further in the chapter, but lower natural gas prices have allowed U.S. refineries to outcompete many European and Island-based refineries, which have to rely on oil to meet their energy and electricity needs at a refinery.

To add even more complexity, the naming convention for hydrocarbons usually varies from *company* to *company*:

Name	Alkanes	Alkenes	Cycloalkane	Arene
Other Names	Paraffins	Olefins	Naphthenes	Aromatics
Example Hydrocarbons	Methane, Pentane	Ethylene, Butylene	Cyclopentane, cyclohexane	Benzene, Xylene
Type of Chemical Bond	Single Bond	At least 1 double bond	Single bond/ring in a chain/ring structure	At least 1 double bond, arranged in a ring/chain structure
Exists naturally in crude oil?	Yes	No, can only be produced from a "cracking" process	Yes	Yes
Common uses as a purity product	Used as a fuel, feedstock for main refinery processes (cracking)	Converting olefins into *poly*-olefins that become the basic building blocks of modern plastics	Primarily as a base building block for advanced plastics (i.e. nylon)	Used in the manufacturing of polymers and used as feedstock for other processes
Typical Gasoline composition by volume	29-48%	2-5%	3-7%	21-54%

What is gasoline?

Gasoline is a mixture of more than 200 *different* liquid hydrocarbons, with the number of carbon atoms ranging from C4 to C12 that are specially blended with additives to meet the needs of the modern combustion engine[91]. The predominant market for gasoline is private vehicles in the transportation sector[92]. The following are important refinery process outputs that are blended together to obtain motor gasoline:

Name	Unit that Produces this output	Characteristics	Typical Percentage of Summer Gasoline
Reformate	Catalytic Reformer	High octane number, low vapor pressure	30%
Cat Cracked Gasoline	Fluid Catalytic Cracker	High content of olefins and medium amount of aromatics	40%
Hydrocrackate	Hydrocracker	Medium to low octane rating, moderate levels of aromatics	5%
Alkylate	Alkylation Unit	High octane blending component. Essentially no olefins, benzene or aromatics. Low Vapor Pressure	15%
Isomerate	Catalytic Isomerization Unit	Free of sulfur, aromatics and olefins	10%

---

[91] http://pubs.acs.org/cen/whatstuff/stuff/8308gasoline.html
[92] Gasoline, in some smaller countries with unreliable electric infrastructure, is also used in a much smaller for backup domestic power generation.

Worldwide gasoline consumption has grown substantially over the past 30 years, going from less than 16MMBPD to over 23MMBPD currently.

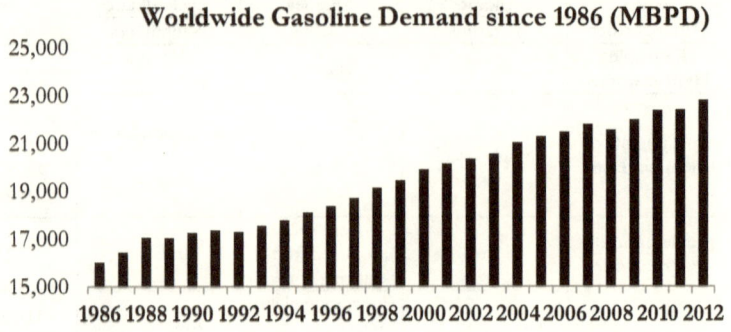

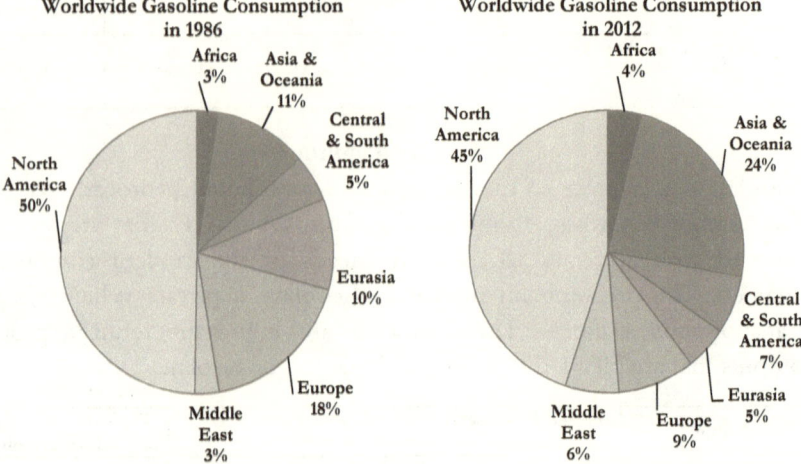

Gasoline demand, particularly in Asia, has increased over the past 30 years, especially since China opened up their economy in the early 1980's. In contrast, Europe's demand for gasoline has decreased by about 30%[93].

What is diesel?

Diesel is a mixture of hydrocarbons with the number of carbon atoms ranging from C12 to C20. Diesel typically has 18%-30% more energy per gallon than regular gasoline, so it is a much more *energy dense* fuel. Diesel is particularly valued in applications where torque, efficiency, durability and lower volatility are valued, such as in the heavy truck market. Diesel is

---

[93] https://www.eia.gov/beta/international/

predominantly used in the heavy duty transportation sector, including railroads, heavy trucks, buses, caterpillar trucks and marine vessels. Diesel also plays an important role in several countries' power generation sector, particularly countries with limited energy resources. Diesel demand is also very regional with areas such as North America and the Middle East demanding more gasoline than diesel[94]:

Region	Diesel/Gasoil Demand as % total Country Demand for Transportation Fuels
European Union	45%
Africa	41%
S&C America	34%
World Average	29%
Asia Pacific	27%
CIS	25%
Middle East	24%
North America	21%

Demand for diesel worldwide has grown substantially more than gasoline:

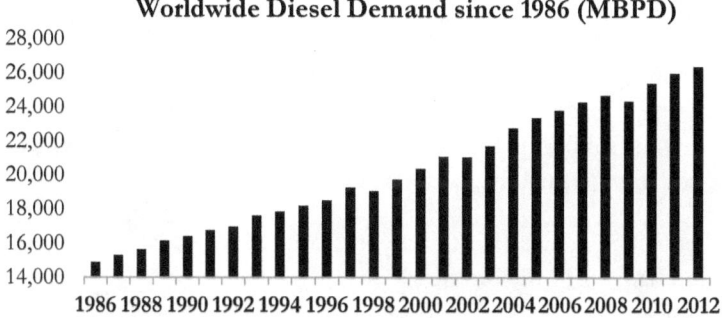

Diesel has grown from less than 15MMBPD in 1986 to more than 27MMBPD currently. Demand in Asia for diesel fuel has grown substantially more than total world demand going from less than 3MMBPD in 1986 to more than 9MMBPD today, a threefold increase in 30 years. In contrast, European demand has only grown by about 25% in the same period.

---

[94] BP Statistical Review 2016, percentages for 2015

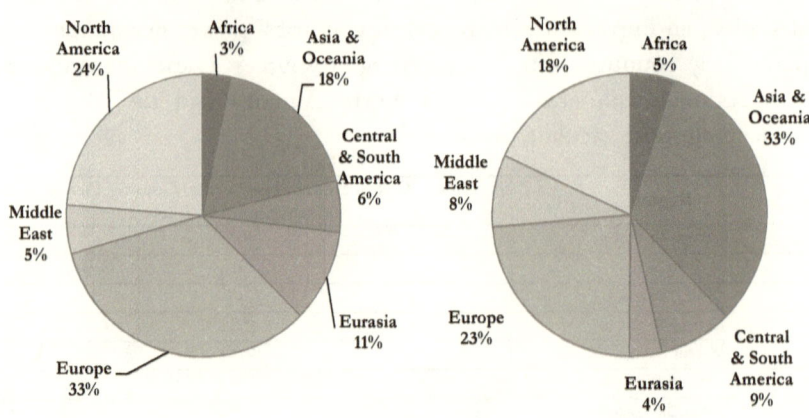

Worldwide Diesel Demand in 1986

Worldwide Diesel Demand in 2012

What is jet fuel?

Jet fuel is another type of distillate fuel that has between 8 and 16 carbon atoms. There are two major type of jet fuels used by commercial and civil aviation, Jet A and Jet A-1[95]. Jet A, which is mainly used in the United States, has a freeze point of *minus* 40° Celsius while Jet A-1 must have a freeze point of *minus* 47° C. Millions of gallons of jet fuel are put into turbine-powered aircraft every day. Jet fuel is a middle distillate that is produced with similar processes to diesel, but has very specific fuel additives added to jet fuel that address several areas:

- Protection against frost and corrosion, which cause integrity issues and damage to the turbines.
- Biocides to prevent bacteria from growing. Certain bacteria and fungi are capable of existing in the water where it interfaces with the fuel. These microorganisms use alkanes and additives in fuel as foodstuff. These microbes can propagate rapidly. The by-product is a sludge-like substance. In sufficient quantity, this substance can cause corrosion on steel and aluminum surfaces and attack rubber fuel system components[96].
- Reduction of soot or particle emissions from combustion of this particular fuel.

---

[95] https://www.exxonmobil.com/en/aviation/products-and-services/Products/Jet-A-Jet-A-1
[96] http://www.aviationpros.com/article/10387588/fuel-contamination-increasing-awareness-on-factors-that-lead-to-jet-fuel-contamination

- Increase of lubricity or the reduction of friction or wear by the lubricants that are added to an aircraft.
- Jet fuel needs to be specifically free of water; therefore water moisture repellants are added. Water can promote corrosion in fuel system components. If enough water is present, it can form ice crystals in low temperatures and clog fuel lines, filters, or components. This could disturb or even stop the fuel supply to the engine[97].

Although jet fuel is a commodity that is bought and sold in terms of volumes, it must have the following properties[98]:

- Stability, whereby jet fuel's properties remain unchanged *despite* changes in temperature or pressure.
- Lubricity, a measure of a material's effectiveness as a lubricant. Jet fuel must possess a certain degree of lubricity because jet engines rely on the fuel to lubricate some parts in fuel pumps and flow control units.
- Fluidity, the ability of the jet fuel to flow freely from fuel tanks in the wings to the engine through an aircraft's fuel system, which includes the *viscosity* and *freezing point* of the fuel.
- Volatility, which is a fuel's tendency to vaporize, with jet fuel being fairly stable.
- Non-corrosive, it is essential that the jet fuel does not corrode any of the materials that it comes in contact with during distribution and use.
- Cleanliness, which indicates the absence of solid particulates and free of water.

Jet fuel demand has grown by about 50% over the past couple of years in the world particularly in Asia. This massive increase is despite significant fuel efficiency improvements developed in the field of jet engines and aerodynamics, resulting in higher jet engine fuel efficiency. Jet fuel burned per seat is 70% *less than* that of the early jet engine era in the 1960's[99].

---

[97] Ibid
[98] https://www.cgabusinessdesk.com/document/aviation_tech_review.pdf
[99] https://www.transportenvironment.org/sites/te/files/media/2005-12_nlr_aviation_fuel_efficiency.pdf

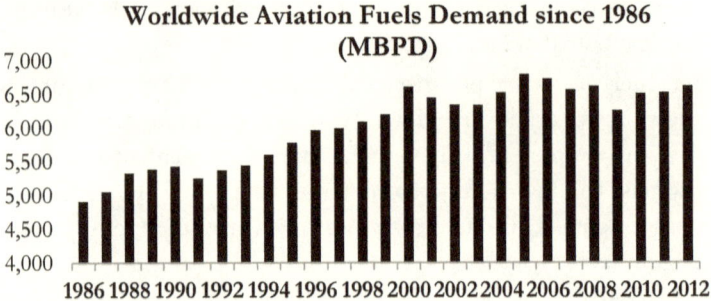

Worldwide Aviation Fuels Demand since 1986 (MBPD)

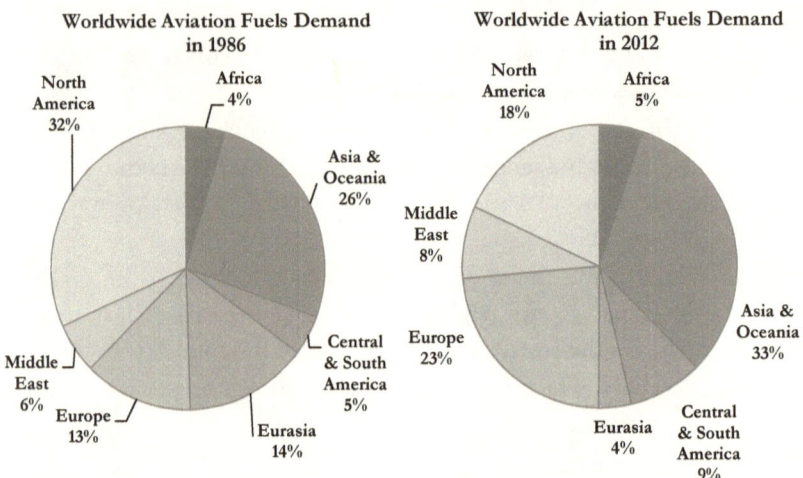

Worldwide Aviation Fuels Demand in 1986

North America 32%
Africa 4%
Asia & Oceania 26%
Central & South America 5%
Eurasia 14%
Europe 13%
Middle East 6%

Worldwide Aviation Fuels Demand in 2012

North America 18%
Africa 5%
Middle East 8%
Asia & Oceania 33%
Central & South America 9%
Eurasia 4%
Europe 23%

Octane Rating

The octane rating measures the ability of a fuel to burn in a *controlled manner* and at the *desired time*. The octane rating measures the ability of a fuel to burn *in comparison* to the iso-octane molecule, with iso-octane being assigned a rating of 100. This is the reason why this rating is called *octane rating[100]*. In other words, gasoline with an octane rating of 91 would have to combust in the *same manner* as a mixture of consisting of 91% iso-octane molecule. It is important to note that octane rating *does not relate* to the actual energy content of the fuel being considered, it is only a measure of the fuel's tendency to burn in a certain manner[101]. The *higher* the octane number, the *higher* the resistance of that fuel to ignite prematurely, thus preventing

---

[100] http://www.pei.org/wiki/octane-number
[101] Ibid

knocking[102] in the engine. Left untreated, engine knocking can cause damage to the engine; prevent the vehicle from operating correctly and in some high-end vehicles that require high octane gasoline may even void the manufacturer's warranty[103].

Octane rating is measured by two numbers, the RON, or Research Octane Number and MON or Motor Octane number[104]. These two numbers are then averaged out to arrive at the published octane ratings that we see at the fuel pumps in every service station.

Overall, engines with *low number* of cylinders would tend to have *higher* compression ratios and thus require *higher octane* rating gasoline than bigger cylinder engines[105]. As fuel economy increases, higher compression engines will be required, over time increasing the demand for *higher* octane gasoline. This increased demand for *higher* octane fuels presents a challenge for refiners as they need to increase the production of high octane gasolines and incorporate other hydrocarbons or alcohols, like iso-butane or ethanol, respectively, which have higher octane ratings than conventional gasolines.

Cetane Number

In contract to the octane rating, the cetane number is used in diesel fuels. Cetane number is a measure of the fuel's delay of ignition time, in other words, the amount of time between the *injection* of fuel into the combustion chamber and the *actual* start of combustion of the fuel charge. Cetane value also relates to how *well* the diesel engine starts in cold weather. There is no spark plug to fire up ignition, so the diesel engine has to have the fuel be warmed up in temperature before it will ignite. Diesel fuel with a higher cetane fuel ignites more easily and makes the diesel easier to start in cold weather[106].

Reid Vapor Pressure

The Reid Vapor Pressure, or RVP for short, is a way of measuring the *volatility* or *evaporation* characteristics of liquid fuels. Volatility is a measure of how easily a liquid (or solid) will change into a vapor state. In other words,

---

[102] Engine knock, or pinging, occurs when a separate pocket of air-fuel mixture ignites after the spark has ignited the air-fuel mixture within the combustion chamber. Source: Engine knock, or pinging, occurs when a separate pocket of air-fuel mixture ignites after the spark has ignited the air-fuel mixture within the combustion chamber

[103] https://www.cars.com/articles/if-premium-gas-is-recommended-for-my-car-will-using-regular-void-the-warranty-ruin-the-engine-1420684149253/

[104] http://inside.mines.edu/~jjechura/Refining/02_Feedstocks_&_Products.pdf

[105] https://www.carthrottle.com/post/engineering-explained-high-vs-low-octane-petrol/

[106] https://www.bellperformance.com/blog/bid/102350/What-Cetane-Value-Does-in-Diesel-Engines

the vapor pressure measures the *tendency* of a liquid to turn into *vapor*. A liquid with a *higher* vapor pressure will turn into vapor at a *lower* temperature. The vapor pressure of *lighter* hydrocarbons, such as ethane or propane, is *higher* than the vapor pressure of *heavier* hydrocarbon groups, such as naphtha or distillates.

In the United States, the Environmental Protection Agency or EPA regulates the vapor pressure of gasoline sold at stations across the country to minimize evaporative emissions[107]. Since the volatility of any liquid depends on the surrounding environment's temperature, liquids in warmer climates will have a tendency to vaporize *quicker* than in colder climates.

Typically, summer-grade gasolines have a *lower* volatility than winter-grade gasolines. This change in RVP for summer and winter gasoline blends is done in order to limit evaporative emissions that increase with warmer weather[108]. All gasoline blends in the US have to be below 14.65PSI, which is atmospheric pressure, but the summer blends have to be typically below 9PSI or 7.8PSI, depending on the area[109]. Winter blends are typically above 9PSI, but below the required 14.65PSI RVP.

Why do we need Refining?

The refining problem can be succinctly summarized as follows:

> *"What the world desires in refined petroleum products does not exist in natural conditions"*[110]

The main objective of refining is to convert a largely unusable commodity (oil and feedstocks) in their *natural state* into valuable products that consumers demand such as transportation fuels. We could not be able to simply use crude oil in our modern day combustion engines since fuels need to have certain properties in order to fully and successfully combust. In other words, consumers demand 100 barrels of gasoline, but only 30 barrels of gasoline type molecules exist *naturally* in crude oil. The overall goal of refining is to successfully and profitably meet what end-users require with today's available hydrocarbons.

---

[107] https://www.epa.gov/gasoline-standards/gasoline-reid-vapor-pressure
[108] https://www.eia.gov/todayinenergy/detail.php?id=11031#
[109] https://www.epa.gov/gasoline-standards/gasoline-reid-vapor-pressure
[110] Unknown author

The following table provides an illustration of two types of crude oil, one light and one heavy, as well breaks down worldwide demand of petroleum products by group[111]:

Group	Brent Crude[112] Percent	Heavy Crude Percent[113]	Worldwide Demand Percent of Total	Worldwide Demand (MMBPD)	Typical Products	Biggest consumers
Light Distillates	27.1%	15.1%	32.5%	30.8	Aviation & motor gasolines, light distillates feedstock	Transportation sector
Medium Distillates	31.9%	18.6%	36.4%	34.6	Diesel/gasoil fuel, jet fuel, gasoline blend stock heating oil, marine bunker fuel	Transportation sector
Fuel Oil	27.1%	31.5%	8.4%	8.0	Marine bunker fuel, crude oil used directly as fuel, residual fuel	Marine transportation, power generation
Other Products	13.9%	34.8%	22.8%	21.6	Refinery gas, LPG, petroleum coke, lubricants, bitumen, waxes, other products and fuel losses	Various, including plastics, lubricants, Petchem uses, aluminum production and many others
Total	100%	100%	100%	95MMBPD	-	-

These molecule groups do not *naturally* exist in quantities sufficient enough to meet demand in their *natural state* in crude oil. The other consideration is that refined products themselves have different specifications for different properties, such as octane or cetane ratings, vapor pressure, emission controls, and other requirements that are simply *not found* in crude oils in their natural state.

As these molecule groups do not naturally exist in their final form in crude oil, this is why refining is needed. For example, the world typically demands around 32-33% of all refined petroleum products to be in the form of light distillates (primarily motor gasoline), but a light crude oil like Brent only contains *naturally* about 27% in light distillate type molecules. When compared with a heavy crude oil like Cold Blend, only 15% of that crude oil's naturally occurring molecules are in the light distillate category. The primary objective of a refinery is to *distill, transform, convert* and *upgrade* crude

---

[111] 2016 BP Statistical Review, Oil Regional Consumption table
[112] Brent Crude here is used an example. Brent is a reference crude for pricing crude oils around the world. Crude Assay from ExxonMobil http://corporate.exxonmobil.com/en/company/worldwide-operations/crude-oils/brent-blend
[113] Canadian Cold Lake Blend. Crude Assay from ExxonMobil
http://corporate.exxonmobil.com/en/company/worldwide-operations/crude-oils/cold-lake-blend

oil's naturally occurring molecules into the desired molecule groups that the world requires.

The table below provides an illustration of how much crude oils can vary from one another[114]:

Category	Brent Crude	Basra Heavy	Qua Iboe	Mondo	Terengganu	Ebok
Area/Region	North Sea	Iraq	Nigeria	Angola	Malaysia	Nigeria
API Gravity	37.6	24.0	37.6	27.5	73.1	19.0
% Sulfur	0.404%	3.83%	0.106%	0.438%	0.002%	0.397%
Composition by type of hydrocarbons						
Alkanes	36.5%	33.7%	34.2%	22.9%	73.1%	6.4%
Cycloalkanes	34.9%	17.3%	42.6%	36.3%	18.4%	43.3%
Aromatics	28.6%	49.0%	23.2%	40.8%	8.5%	50.3%
Total	**100%**	**100%**	**100%**	**100%**	**100%**	**100%**
Composition by group of molecules						
Light Distillates	27.1%	15.9%	27.1%	16.8%	90.2%	5.6%
Medium Distillates	31.9%	25.7%	39.9%	29.6%	8.3%	36.8%
Fuel Oil	27.1%	29.3%	25.8%	30%	0.3%	37.6%
Other Products	13.9%	29.1%	7.2%	23.6%	1.3%	20.1%
Total	**100%**	**100%**	**100%**	**100%**	**100%**	**100%**

An additional factor not necessarily captured by simply analyzing the molecule composition is how fuel requirements have changed over the last 100 years or so. The first oil refineries in the late 19th century are certainly not the same as today's modern and highly complex refineries. Since the construction of the first refinery, significant changes have occurred to the refining business environment and value chain, with some examples being:

- Before the invention of the modern combustion engine and before mass electricity was available, the main goal of a refinery was to produce as much kerosene as possible, which was highly demanded since it was used for lighting[115]. In fact, those early refineries were designed to *maximize* kerosene or middle to heavy distillates yield while *minimizing* production of any other fractions, even to the point of letting evaporate or go to waste any fractions with a boiling point lower than the kerosene[116].

- After electricity became *widely* available in households across the United States and Europe, demand for kerosene for illumination

---

[114] http://corporate.exxonmobil.com/en/company/worldwide-operations/crude-oils/crude-blends-by-characteristic

[115] Source: The Prize, by Daniel Yergin, page 63

[116] http://www.chemistryexplained.com/Fe-Ge/Gasoline.html

purposes began to decline substantially, since the illumination market was more efficiently served by electricity rather than kerosene. Threatened by declining demand for their largest single product, refineries started to look for new markets to sell the refined petroleum products to.

- The answer to the search for new markets came from the mass production and sale of automobiles, which caused demand for gasoline to grow exponentially. Refineries now had to find *new ways* to increase the production of gasoline, which led to many improvements in the refining process, including the invention of catalytic cracking[117], and other processes which were aimed at converting *more* and *more* molecules into gasoline type ranges.

- Beginning in the middle to late twentieth century, increased environmental regulations led to phasing out gasoline components such as tetraethyl lead, methyl tertiary butyl ether (MTBE), benzene and other components that were determined to be harmful to the environment[118]. In recent years, environmental regulations have increased leading to more stringent specifications for refined products. This has caused refineries to increase capital investments aimed at *reducing* these components from finished products. Recent examples, like the push for ultra-low sulfur diesel and gasoline have impacted refineries by requiring more processing equipment and sulfur removal from these fuels[119].

- In the United States, federal, state and local governments have a significant impact into fuels regulation. These regulations have increased substantially over the past 20 years or so. As early as 1990, only six types of gasoline were sold in the US. Today, there are more than 15 *unique* gasoline formulations (not including the different octane ratings of regular, mid-grade and premium) and each gasoline formula is manufactured for specific markets throughout the country[120].

- Over the past year, the International Maritime Organization, a UN agency, has set new standards to reduce the maximum allowable sulfur content for marine vessels from 3.5% to 0.5% by 2020,

[117] http://www.ogj.com/articles/print/volume-90/issue-20/in-this-issue/general-interest/fluid-catalytic-cracking-hits-50-year-mark-on-the-run.html
[118] http://www.eesi.org/papers/view/fact-sheet-a-brief-history-of-octane
[119] http://www.hydrocarbonprocessing.com/magazine/2016/february-2016/trends-and-resources/business-trends-clean-fuels-a-global-shift-to-a-low-sulfur-world
[120] http://www.nacsonline.com/YourBusiness/FuelsCenter/Alternative/Articles/Pages/Boutique-Fuel-Map.aspx#.WQ6KAn1tmdQ

causing refineries across the world to have to make substantial changes into how they process residual bunker fuel in the refinery and produce other products[121]. Refineries will have to make substantial investments to either process further residual fuel, add more desulfurization capacity or be forced to sell bunker fuel to developing countries for uses like power generation[122].

Modern refinery operations are indeed very complex due to the vast array of feedstock sources, qualities and sophisticated processing technology required. What increases the complexity is the fact that processing units and products produced within the refinery are interrelated. Understanding the unit and product dependencies in a refinery is critical in understanding *how* a refinery makes money.

Refining Process

The refining process can be thought of being comprised of four key functions[123], *separation, contaminant removal, conversion and blending.*

- Separation: Consists of separating the individual components or *groups* of components into what are called "cuts" or "fractions". There are two major types of distillation in a refinery[124], *atmospheric distillation* and *vacuum* distillation.

- Contaminant removal: There are several process units in a refinery dedicated to removing certain contaminants, but among those the following are the most significant:
    - Desalting, removing salt and other sediments from crude oil before it is distilled.
    - Hydro treating, which is a process that uses hydrogen and a catalyst to remove sulfur in order meet product specifications.
    - Catalytic oxidation, which also uses a catalyst to remove certain sulfuric compounds from crude oil and other hydrocarbons.

---

[121] http://www.platts.com/IM.Platts.Content/InsightAnalysis/IndustrySolutionPapers/SR-IMO-2020-Global-sulfur-cap-102016.pdf
[122] https://www.eia.gov/todayinenergy/detail.php?id=28952
[123] For a more detailed analysis of refining processes a good resource is Dr. Bill Leffler's "Petroleum Refining in Non-Technical Language", published by PennWell
[124] A refinery may also have an NGL fractionator, which uses *cryogenic* fractionation technology

- Conversion which includes processes such as:
  - Cracking or breaking of *larger* hydrocarbon molecules into *smaller* hydrocarbon molecules.
  - Alkylation and polymerization, which involves combining *smaller* hydrocarbon molecules into *larger* molecules.
  - Reforming and isomerization, which involves *changing* the *shape* of molecules into more valuable molecules with different characteristics (i.e. higher octane).
- Blending, which consists of *physically* combining different molecules into a new product, an example being gasoline blending where certain components that are produced at the refinery (i.e. alkylate, which has higher octane than refinery gasoline) into transportation fuels to modify certain desirable properties, such as octane rating, vapor pressure or other.

Crude Distillation

Crude distillation is the first step in the refining process after crude desalting. Distillation is the process of *separating* components from a liquid mixture by selective *evaporation* and *condensation*. Refineries are usually measured in terms of atmospheric distillation capacity and distillation is the most basic unit for crude oil processing. Unlike the other processes in a refinery, distillation does not involve *chemistry* per se in the sense that distillation is simply a physical separation process, in other words the *same* molecules in the crude oil going through the distillation unit are the *same* molecules coming out from the unit, the only difference is that they are *physically* separated into different *fractions* or *cuts*. Historically, distillation is by far the oldest refining process. The overall goal of distillation is to separate the crude oil into fractions with different boiling point. The following table provides an indication of the typical boiling point ranges for several hydrocarbons[125]:

Product	Boiling Point Range	Number of Carbon Atoms
Butane and lighter products	<85° F	Less than 4 (<C4)
Gasoline blending components	85°-185° F	5-10 (C5-C8)
Naphtha	185°-350° F	(C8-C12)
Kerosene, jet fuel	350°-450° F	(C12-C18)
Distillate (diesel, heating oil)	450°-650° F	(C18-C24)
Heavy Gas Oil	650°-1050° F	(C24-C30)
Residual Fuel Oil	<1,050° F	C30+

---

[125] https://www.eia.gov/todayinenergy/detail.php?id=6970

As shown on the prior table, lighter hydrocarbons have a *lower* boiling point, meaning that they will go from a *liquid* state to a *gas* state *above* these boiling points. In general, as crude oil is heated and goes through a distillation tower, the lighter components will vaporize *first* and rise through the different columns and condense at different levels, depending on the temperature. Each fraction will be sent for further processing at different units throughout the refinery. The following diagram provides an overview of all the different intermediate products that are distilled from crude oil[126]:

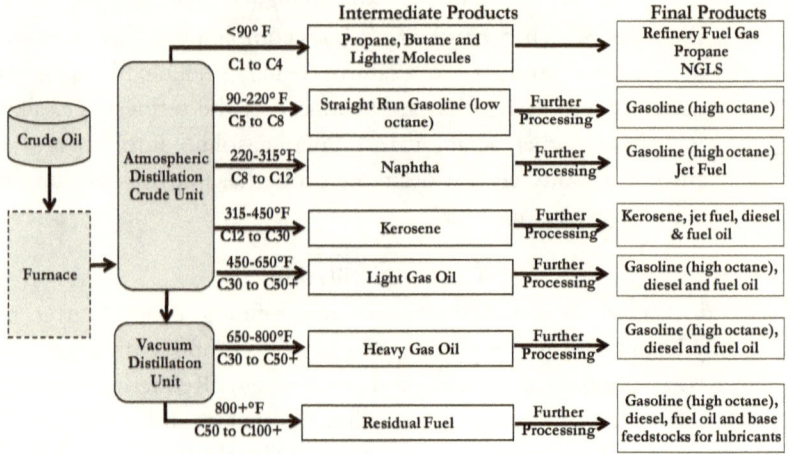

There are several distillation methods:

- Atmospheric Distillation: Main type of distillation used at a refinery and the capacity of these distillation units serves to rank refineries in terms of size. As the name implies, atmospheric distillation is completed at *atmospheric* pressures, in contrast to *vacuum* distillation. Atmospheric distillation is usually accomplished by several towers which have different *trays* for distilling each particular cut or fraction. These cuts are then sent for further processing at different units at the refinery. Atmospheric distillation can be used for all types of crude oils since it is the very first step in crude oil refining.

- Vacuum Distillation: As it name implies, vacuum distillation accomplishes distillation by being completed in a *vacuum* or in a *no pressure* system (0.4 psi vs. 14.65 psi in ambient conditions) and thus be able to distill *heavier* residues from crude oils. By reducing

---

[126] Valero Corporation's Refining 101, January 17, 2013 Presentation, slide 7

the temperature, cuts with higher boiling points can be distilled at *lower pressure* and thus prevent unwanted *cracking* or breaking of larger molecules.

- Cryogenic Distillation: *Cryogenic* means *very low* temperatures, several degrees below zero Celsius. Cryogenic distillation is particularly useful in separating natural gas liquids which have very low boiling points, an example being ethane, where its boiling point at atmospheric pressure is *negative* 127° F[127].

Contaminant Removal

Several processes are involved in contaminant removal:

Process	High Level Description	Background	Output
Desalting	Removes salt, dirt and other sediments from raw crude oil before it is distilled in a refinery	First process in refining. Improves processing by preventing corrosion in equipment throughout the refinery	Unsalted crude oil
Catalytic Oxidation	Process that "burns off" certain unwanted compounds using catalysts.	Helps meet product specifications by removing certain organic compounds, like mercaptans (sulfuric compound)	"Sweetened" crude fractions or gases. Byproducts are typically sulfur compounds.
Hydrotreating	Removes contaminants like sulfur, nitrogen, oxygen and metals	Uses hydrogen and catalysts to remove unwanted contaminants like sulfur, nitrogen and other components. Consumes significant amounts of hydrogen, which in some refinery locations may present challenges as to where to source hydrogen	Cleaner refinery products and feedstocks. As by-products, sulfur and other contaminants removed

Conversion Processes

Cracking

Cracking is the process of transforming or breaking larger molecules into smaller *more valuable* molecules. For example, gasoline is usually composed of hydrocarbons with 5 carbons to 12 carbons, while crude oil molecules are usually much higher in carbon atoms. The goal of cracking is to turn these *longer chain* molecules that are not demanded by the market and to convert them into more desirable *shorter chain* molecules or other intermediate feedstocks.

Cracking usually involves using a combination of heat, pressure and catalysts to accelerate the breaking of these molecules. Thermal cracking on

---

[127] http://www.airproducts.com/products/Gases/gas-facts/physical-properties/physical-properties-ethane.aspx

the contrary is the only cracking process that relies only on increases in temperatures and pressures thus not requiring any catalyst for the reaction to take place.

Cracking encompasses several different processes:

Process	High Level Description	Background	Typical Products[128]
Thermal Cracking	Uses high temperatures and pressures to crack molecules	First cracking method invented in the late 1800's	Typically residual fuels & light oils
Catalytic Cracking or FCC	Uses a catalyst, typically silica, alumina, zeolites and other solid acids in fluidized form, to *accelerate* cracking under lower temperature and pressures than without a catalyst	Developed in the 1940's to increase gasoline and diesel yields while minimizing heavy oil production. Most refineries in the US have an FCC unit	Light gases & olefins (propylene & butylene), used as petrochemical feedstocks. Light & Heavy naphtha & cycle oils, used in transportation fuels.
Hydrocracking	Hydrocracking uses high pressures, large amounts of hydrogen and catalysts in order to increase yields of distillates and reduce production of heavy fuel oils. Hydrocrackers tend to be more costly to install than FCC's	Developed in the 1960's to increase production of transportation fuels. More commonly installed in European refineries due to higher local demand for diesel	Produces high yield of clean products, in particular diesel, since hydrogen suppresses coke formation. Enables capture of arbitrage between natural gas and crude oil prices
Coking	Coking uses high temperatures to "cook" or crack very long chain molecules, commonly known as "bottom of the barrel" and extract as much as possible and thus reduce unwanted residual fuel production	Cokers are the hallmark units of very complex refineries and are used to process *heavy* and *extra heavy* crude oils and increase production of gasoline and diesel	Light gases, distillates with high content of sulfur and olefins as well as petroleum coke
Visbreaking	Similar to thermal cracking, it works by reducing the viscosity of the feeds. It does not use a catalyst and it is less expensive than a thermal cracker	Developed in the 1930's to produce more desirable heating oil (at the time) and other products and reduce the viscosity of residual fuel[129]	Typically low yields of high value products and high yields of fuels like residual fuels, which need to be further cracked, particularly in a hydrocracker

Alkylation & Polymerization

The primary role of alkylation is to re-combine some of the smaller molecules (olefins or alkenes like propylene[130] and butylene) produced by the cracking processes, specifically, the FCC, into larger molecules in the preferred gasoline range[131]. One of the key advantages of producing *alkylate* is that the re-combined molecules have a *higher* octane rating. Feeds into the

---

[128] Source: http://inside.mines.edu/~jjechura/Refining/
[129] http://www.sciencedirect.com/science/article/pii/S1026309812000612
[130] Olefins or Alkenes are hydrocarbons that have at least one double-bond and are highly reactive. For example propylene has three carbons and 6 hydrogens, with one double-bond and one single-bond.
[131]http://www2.emersonprocess.com/siteadmincenter/PM%20Micro%20Motion%20Documents/Refining-Alkylation-PSG-MC-00935.pdf

alkylation unit are butylene which is then combined typically with an NGL like iso-butane to form iso-octane, which is a high-octane blending component.

Process	High Level Description	Background	Typical Inputs/Outputs
Alkylation	Combines olefins like butylene with an NGL like iso-butane to produce high-octane gasoline (iso-octane molecule)	First developed in the 1930's to improve the octane rating of aviation gasoline. This invention was critical in assisting the war effort in the 1940's[132].	Olefins from the FCC combined with NGLs from light gases to produce iso-octane
Polymerization	Combines a light olefin like propylene, which is reacted with more propylene to form a hexane, which is also a higher octane gasoline blending component	First developed in the 1950's, it has been largely phased out and most refineries prefer the alkylation method[133]	Olefins from the FCC combined to form alpha olefins like hexene, which can then be blended into gasoline to improve the octane rating

Reforming & Isomerization

The overall goal of reforming and isomerization is to change the shape or the chemical *structure* of the molecules going through these processes in order to improve, typically, the octane rating and other qualities of the products being transformed. For example, *normal* butane, chemical formula C_4H_{10}, has different boiling points, vapor pressure and different energy content than *iso*[134]-butane[135], having the same chemical formula C_4H_{10}, but a different chemical *structure*.

Process	High Level Description	Background	Typical Inputs/Outputs
Reforming	Reforming transforms (cycloalkanes) naphthenes molecules like heavy naphthas to aromatics. Produces hydrogen that can be used in other refining processes	Invented during World War II to produce toluene which was used to produce dynamite for the war effort[136]. Also produces benzene, which has to be removed and is used as a petrochemical feedstock[137].	Heavy naphthas reformed with certain catalysts to produce, "reformate" which is a high octane feedstock with *low* vapor pressure and *low* sulfur levels.
Isomerization	Converts alkanes like normal butane into *isomers* like isobutane, which increase the octane rating and provide other benefits	Developed as well during WWII to improve the octane rating of aviation gasoline.	Typically light naphthas, normal butane and hydrogen processed with catalysts to produce high octane components.

[132] http://inside.mines.edu/~jjechura/Refining/09_Gasoline_Upgrading.pdf
[133] Ibid
[134] An *isomer* is a molecule with the same chemical formula but with a *different* chemical structure
[135] https://rbnenergy.com/you-can-just-iso-my-butane-isobutane-and-isomerization
[136] http://inside.mines.edu/~jjechura/Refining/09_Gasoline_Upgrading.pdf
[137] https://emergency.cdc.gov/agent/benzene/basics/facts.asp

The reformer is a major gasoline-producing unit, providing about one-third of the gasoline volume a refinery produces. A reformer takes low-octane gasoline material and "reforms" molecules to produce molecules with more complex structures and higher octane than the simpler naphtha feedstocks[138]. Catalytic reforming produces what is commonly called *reformate* which is a critical component of gasoline blending and accounts for about 30% of gasoline supplies in the United States[139].

Volumetric Gain/Loss by Refining Process Unit

Volumetric or processing gain refining is defined as the following:

> *"The volumetric amount by which total output is greater than input for a given period of time. This difference is due to the processing of crude oil into products which, in total, have a lower specific gravity than the crude oil processed.[140]"*

Since liquid hydrocarbons like crude oil and refined products are bought and sold by volume, there will always be a recorded "gain" or "loss". Recorded volume, but not *mass* or *weight*, is changed per se when crude oil is processed in a unit.

The processing gain or loss depends heavily on the types of units a refinery has as well as the type of crude oil being processed. Typically, *heavier* crude oils would experience a *higher* processing gain than *lighter* crude oils due to the fact that heavier crude oils have a higher proportion of *higher molecular density* hydrocarbons. The following chart showcases monthly refinery processing gains in the U.S. since 2005[141]:

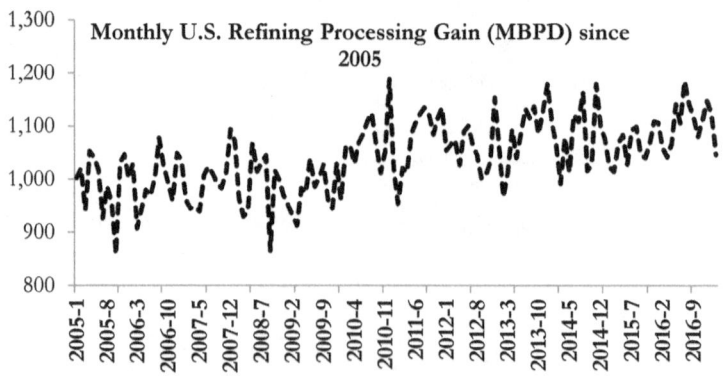

---

[138]https://www.eia.gov/pub/oil_gas/petroleum/feature_articles/2010/distillateprod/distillateprod.pdf?src=email

[139] https://rbnenergy.com/don-t-fear-the-catalytic-reformer-changes-in-the-us-reformate-market

[140] https://www.eia.gov/dnav/pet/TblDefs/pet_pnp_refp2_tbldef2.asp

[141] https://www.eia.gov/dnav/pet/hist/LeafHandler.ashx?n=pet&s=mpgrx_nus_1&f=a

The following table summarizes the typical volume gain/loss by refining unit:

Refining Unit	Volume Gain/Loss on Feedstock to Unit[142]	Why?
Reformer (running at high severity)	*Losses* of -30% to -20%	Produces hydrogen and aromatics as by-products, which reduce volume out of the produced reformate (high octane blending component)
Reformer (running at low severity)	*Losses* of -20% to -10%	Produces hydrogen and aromatics as by-products, which reduce volume out of the produced reformate (high octane blending component)
Fluid Catalytic Cracker	*Gains* of +5% to 15%	Volume is gained due the *cracking* of longer chain hydrocarbons into shorter chain hydrocarbons, which have a *lower density*, thus occupying more volume
Alkylation	*Losses* of -20% to -25%	By combining smaller molecules (like isobutane and butylene) which *occupy more space* into larger molecules (isooctane) which take up less space, measured output is *less* than the input
Coker	*Losses* of -25% to -35%	In all coking units there is a percentage of residual fuel oil that will form coke and since coke is a solid that occupies *less space* than residual fuel oil, measured volume is *lost*
Full-conversion Hydrocracker	Gains of 20% (targeting heavier diesel products) to 35% (targeting lighter gasoline products)	Volume is gained due to the *cracking* of longer chain hydrocarbons into shorter chain hydrocarbons, which have a *lower density*, thus occupying more volume

How do natural gas prices impact refining & downstream?

Natural gas impacts the downstream sector in two main ways, first as an *input* to energy production and second as a *feedstock* to several processes.

Natural gas can be used in a refinery in the following ways[143]:

- Fuel for process and utility heaters, replacing fuel oil.
- Feed and fuel for the hydrogen generation unit, replacing naphtha.
- Fuel for gas turbines, replacing naphtha.
- Fuel for process heaters, replacing fuel gas.

Petroleum refining is a highly *energy-sensitive* industry with energy usage accounting for roughly 50% of cash operating costs[144]. Having access to

---

[142] Valero Refining 101 dated January 17, 2013, slide 18
[143]http://www.digitalrefining.com/article/1000557,Natural_gas_fuels_the_integration_of_refining_and _petrochemicals_.html#.WSr7Yc5tmdQ

inexpensive and abundant natural gas can provide many benefits to refining and downstream in general:

- Lower natural gas prices provide overall lower electricity costs for a refinery, whether that electricity is purchased from third parties using natural gas as a fuel or generated inside the refinery.

- Many processes require large amounts of hydrogen, such as hydrotreating and hydrocracking. Hydrogen can be most cost *effectively* produced by using natural gas[145].

- *Associated* natural gas liquids are produced along with natural gas in upstream wells, which in a case of abundant natural gas production will tend to cause low NGL prices. NGLs are used as feedstocks that a refinery needs such as iso-butane to produce alkylate.

- The chemicals business is even more markedly impacted by low natural gas prices. Thanks to higher NGL production, particularly ethane, chemical plants can produce ethylene at lower costs than naphtha-based crackers by using lower cost ethane[146]. Ethylene cash costs are currently in the U.S. about *less* than $350 per metric ton, while naphtha-based crackers in Asia have ethylene cash costs close to $700 per metric ton[147].

- Ethylene, the most widely used petrochemical feedstock, can be cracked from *both* ethane and higher molecule hydrocarbons like naphtha. In a low natural gas to crude oil price ratio scenario, naphtha is typically the marginal or *highest cost* feedstock, setting worldwide ethylene prices in a tight ethylene market. Ethylene producers with ethane-based crackers can capture *higher margins* by being the *lowest cost* producer of this commodity and selling ethylene at *naphtha-based* prices.

Valero, the world's largest independent refinery has the majority of their refineries located in the U.S. Because of this, Valero realizes a pre-tax cost advantage over European refineries of about $1 billion per year or about $1.07 per bbl[148]. Although $1 per bbl may not sound like much money, the average 10 year EBITDA per barrel of refining companies in the US is

---

[144]https://www.energystar.gov/sites/default/files/tools/ENERGY_STAR_Guide_Petroleum_Refineri es_20150330.pdf

[145] http://www.fsec.ucf.edu/en/consumer/hydrogen/basics/production.htm

[146] https://www.icis.com/resources/news/2013/02/25/9644127/insight-middle-east-moves-up-global-ethylene-cost-curve/

[147] Chevron 2017 Annual Analyst meeting, Downstream segment, slide 6

[148] Valero Investor Presentation, April 2017 slide 25

around $5 per bbl, therefore a $1 improvement may translate into a 20% *higher* EBITDA per barrel, just on lower energy costs alone.

Natural gas prices in the United States have remained low both in *absolute* terms and in *relative* terms to crude oil prices. These low natural gas prices have provided the United States with an unsurpassed advantage of having *lower* energy costs and abundant feedstocks to use in refining and petrochemical processes.

Types of Refineries

Refineries can range in size from small "topping" refineries that simply distil crude oil into its individual components and process a few *thousand* barrels a day to mega-refineries that can process a few *hundred thousand* barrels per day.

In China for example, a growing type of refineries are called "teapot" because of their small size compared to large state-owned refineries. They have grown significantly since they were allowed to directly import crude oil without having to use the large state owned infrastructure[149]. Russian companies also have a significant presence of small "topping refineries" that behave more simply as distillation towers that separate a few lighter components out of a crude oil mixture[150].

Refineries are further classified into the following groups:

- Simple or Hydro skimming refineries, are refineries that provide very minimum upgrading and usually simply distill crude oil into its individual cuts or group of products. These refineries by their very nature can only process or distill *light sweet* crude oils. These refineries may also have sulfur removal units.

- Complex or Cat Cracking Refineries, these refineries usually have the same units as a simple refinery, plus a Fluid Catalytic Cracking or FCC units that can upgrade heavy gas oils and other medium to heavy hydrocarbons into gasoline and petrochemical feedstocks. These refineries may also have other types of cracking such as hydrocracking units, alkylation plants and gas processing plant

---

[149] https://www.ashurst.com/en/news-and-insights/insights/chinese-teapots-the-game-changer-in-chinas-oil-industry/
[150] The largest of Rosneft's "mini-refineries" processes about 30MBPD. https://www.rosneft.com/business/Downstream/Neftepererabotka/

units[151]. These refineries can process light to medium/heavy crude oils with intermediate content of sulfur.

- Very Complex or Coking Refineries, these refineries have all the units in a complex refinery, plus a coker, which eliminates residual fuel oil production[152]. These refineries can process all types of crude oils, including sour, heavy and extra heavy, which tend to be high in residual fuel molecules and usually produce petroleum coke as a byproduct.

The table[153] below provides the typical U.S. yields from three refineries running a *medium* crude oil:

	Simple	Complex	Very Complex
Gasoline	30%	50%	60%
Jet Fuel	10%	10%	10%
Distillate Fuel	20%	25%	25%
Residual Fuel	35%	10%	-
LPG	-	3%	4%
Coke	-	-	3%
Refinery Fuel	8%	12%	13%
Gain	(3)%	(10)%	(15)%

Roles in a Refining Organization

A refinery is a highly complex undertaking with several different groups being involved in the *day-to-day* operations of a modern refinery:

- Operations: Are in charge of day-to-day operations of the facilities, and include activities such as monitoring equipment run rates, reviewing and changing operating parameters of a processing unit as well as monitoring the *quality* and *quantity* of inputs (crude oil and feedstocks) and outputs (products) of the refinery.

- Maintenance, Instrumentation & Electrical: maintenance personnel who are in charge of *planning, scheduling* and *executing* both *preventive* and *corrective* maintenance at a refinery or associated facilities. Maintenance, I&E employees and contractors work very closely with operations in addressing their concerns and in particular in clearing with operations when completing *Lock-out Tag-out* (LOTO) procedures before executing maintenance tasks[154].

---

[151] Source: Petroleum Refining in Nontechnical Language by William Leffler, page 195

[152] Ibid, page 195

[153] Ibid, page 196

[154] http://www.bp.com/en_us/bp-us/what-we-do/bp-pipelines/merrillville-training-center/lock-out-tag-out-procedure-training.html

- Optimization & Planning: this group works very closely with the supply & trading organization in a downstream company. Their major role is to analyze and model the refinery's processed inputs and refined products production to *maximize* margins to the refinery. In conjunction with other groups, this group of people usually runs a refinery's linear programming or LP model to be able to determine what best crude oils and other inputs to run to maximize certain products' yields based on current market conditions. This group also gets involved with planning and preparing a refinery's opex and capital expenditures plan and proposes projects to optimize a particular section or process units at the refinery. For example, if the long-term fundamental demand for residual fuel oil is heading *lower*, the refinery might be interested in investing in a petroleum coker to reduce production of this unwanted fuel.

- Engineering support: Engineers provide a variety of subject matter expertise and work closely with operations, maintenance and optimization personnel depending on their particular engineering discipline. For example, mechanical engineers would tend to work closely with maintenance to optimize maintenance plans and provide root cause analysis of equipment failures. On the other hand, chemical engineers might be more involved with the optimization group to select the optimal catalyst that maximizes the yield of a preferred refined product based on *current* market conditions.

- Safety: a refinery is a very complex operation with many moving parts and safety is the *number one priority* in any refining company. Today, safety procedures exist for most maintenance and operating activities in order to *prevent* injuries, process upsets or incidents that could put that could put employees, contractors, the community or the actual refinery in danger.

  In fact, the incident rate in the petroleum refining industry is the lowest of many other industries, with a recent study putting the incident rate at 0.9 per 100 full time employees (FTE) in refining, in comparison to air transportation, which is 6.9 per 100 FTE or agriculture at 5 per 100 FTE[155].

---

[155] https://www.afpm.org/refinery_safety_at_a_glance/

The following table provides a comparison of the incident rate between different industries:

Incident Rate per 100 Full Time Employees (BLS)

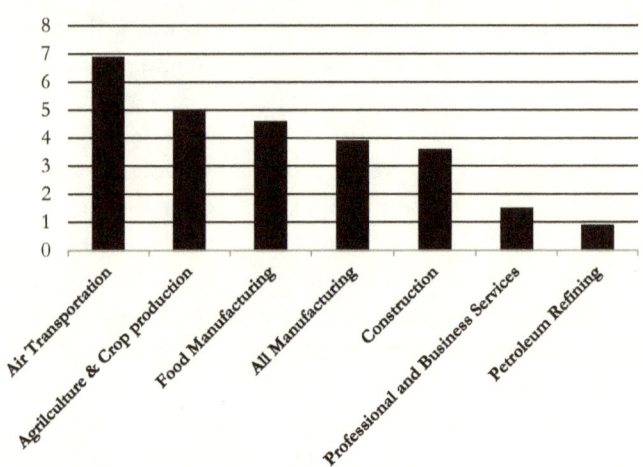

- Turnaround groups: turnarounds, as previously described, are *planned* shutdowns that allow the refinery to complete preventative maintenance or upgrade projects. These planned shutdowns, which can cover the entire refinery or certain process units, allow the refinery to upgrade certain process units, conduct extensive preventive maintenance and otherwise *maintain* and *improve* the safe and reliable operations of the refinery. Most refineries go through a turnaround every three to five years and these turnarounds can last from a few weeks to a few months[156]. Turnarounds require *extensive*, *careful* and *detailed* planning that can precede the actual shutdowns anywhere from a couple of months to a couple of *years*, thus the reason why *permanent* turnaround roles are needed.
- Transactional & Back Office Support: can span several roles and groups within a refinery, from IT operations that might ensure the *distributed control systems* or DCS of the refinery works as expected, to accounting clerks that ensure that all products are accounted for. Transactional accounting roles are key in refining operations since they ensure the critical *charge & yield* processes that allow the refinery to know:
  - o   How much crude oil was processed through the refinery?

---

[156] http://www.apacherefineryservices.com/refinery/

o   How many and what types of products were produced by the different units and in total for the refinery?

o   How many barrels were sold versus kept in inventory?

o   What was the total *volumetric gain* or *loss* at the refinery of converting longer chain molecules such as those found in crude into *shorter* chain molecules, thus yielding a volumetric *gain* or *loss* depending on individual processes at the refinery?

o   Conducting critical *energy & mass* balance at the refinery.

Global Refining Capacity & Throughput

Currently, in terms of atmospheric distillation, the world has refining capacity of approximately 97 million barrels per day (MMBPD)[157]. The top 5 countries in terms of refining capacity are[158]:

- The United States, with approximately 18MMBPD of crude processing capacity, is the largest and most complex refining area in the world. Although no new "greenfield" refinery has been built in the United States since the 1970's, several expansion projects to already existing refineries have increased overall capacity by about 5% in the last 10 years[159].

- China, with 14.2MMBPD of refining capacity, has been growing very fast over the last couple of years, having grown their capacity by about 84% just in the last 10 years alone[160]. Since the economic reforms of the 1980's, when China started to grow at very high rates, the country has added a total of 10MMBPD of refining capacity. China is expected to continue to be a major demand center as current vehicle ownership per capita is relatively low in that country compared to developed economies.

- Russia, with 6.4MMBPD of refining capacity has been growing slightly, with Russia making significant investments in its refineries in order to be able to export fuels to Europe and other markets.

- India, with 4.3MMBPD of refining capacity, their capacity has been growing significantly, with capacity having increased by about 68% just in the last 10 years.

---

[157] 2016 BP Statistical Review, Oil –Refinery capacities table
[158] Ibid
[159] EIA Refining Capacity
[160] Ibid

- Japan, with capacities of 3.7MMBPD, actually has had refining capacity *decrease* by about 18% in the last 10 years alone.

The following table provides an overview of where the world's refining capacity is located at as well as recent utilization rates[161]:

Region	North America	S. & C. America	Europe Eurasia	Middle East	Africa	Asia-Pacific	Total
2015 Total Refinery Throughput (MBPD) 2015	18,976	4,667	19,704	7,390	2,072	26,802	**79,609**
2015 Total Capacity (MBPD)[162]	21,883	6,222	23,635	9,344	3,589	32,554	**97,227**
2015 Utilization Rate	87%	75%	83%	79%	58%	82%	**82%**

For a more detailed table on current capacities by processing unit, please refer to the *Comparative Table number two* at the end of this book.

Refining Hubs around the world

There are several key refining hubs or centers around the world, but the three most significant are the following:

- U.S. Gulf Coast
- Northwest Europe (Rotterdam)
- Singapore

Refining hubs typically have the following characteristics:

- Access to transportation of crude oils and refined products, whether it would be through pipeline, barges or ships.
- Substantial storage capacity, for both crude oil, refined products and other feedstocks.
- Petrochemical facilities or significant access to large consumption centers, whereby feedstocks or by-products can be bought or sold.
- Substantial liquidity from a supply & trading perspective, allowing prices & margins to reflect market fundamentals quicker than at other locations.
- These hubs' margins are usually used as a benchmark for other locations to understand the profitability of the refining business[163].

---

[161] Utilization rates are explained in more detail in the next few pages.
[162] Ibid

U.S. Gulf Coast

The U.S. Gulf Coast is home to approximately 57 refineries, with a total operating capacity of 9.5MMBPD[164], which equates to *roughly half* of US capacity or 10% of the world's refining capacity. The U.S. Gulf Coast refineries are among the most complex in the world being able to process any type of crude oil. This area is also interconnected to many pipeline systems as well as having the capacity to import crude oils, or transport crude oils from inland U.S. production sites, such as the Permian Basin, Eagle Ford and many other areas as well having the terminal capacity to export any refined products. In addition, this area also benefits from nearby access to several petrochemical sites, providing further integration between refining & petrochemicals production.

The U.S. Gulf Coast is also home to the largest hydrocarbon product storage capacity in the world, with current capacities close to 1 billion barrels. Including the Strategic Petroleum Reserve, the storage capacity, as of September 2016, is closer to 1.7 billion barrels[165]:

Commodity	Working Storage Capacity (MBBLs)
Crude Oil	325,978
Gasolines	108,763
Distillates & Jet Fuel	88,320
Residuals, Asphalts & Other Heavies	57,523
Natural Gas Liquids	362,167
Other	118,846
Strategic Petroleum Reserve	727,000
Total	**1,731,071**

Rotterdam

Rotterdam, located in the Netherlands, is a pricing point for refined products in Europe. Over 120 companies have operations in the vicinities of the Rotterdam port. The following assets make up the Rotterdam hub:

- Five refineries encompassing more than 1.2 MMBPD of refining capacity[166].
- Six refinery terminals, with more than 80MMBBLs of storage capacity[167].

[163] https://www.iea.org/media/omrreports/Refining_Margin_Supplement_OMRAUG_12SEP2012.pdf
[164] https://www.eia.gov/petroleum/refinerycapacity/
[165] https://www.eia.gov/petroleum/storagecapacity/ As of September 2016
[166] https://www.portofrotterdam.com/en/cargo-industry/refining-and-chemicals/oil-refineries PDF file, page 16
[167] Ibid, page 18

- Nine refined products terminals, with more than 57MMBBLs of storage capacity[168].
- Forty-two petrochemical sites and sixteen terminals for chemicals and other products.

Singapore

Singapore's strategic location between the Indian and Pacific Oceans and near the Strait of Malacca has allowed it to become one of Asia's major petrochemical and refining centers and oil trading hubs[169]. Singapore has world-class refining, storage, and distribution infrastructure, and Jurong Island on the southern edge of the country is the center of Singapore's petrochemical industry. Several major international energy companies operate facilities in the area. Singapore's government plans to promote long-term growth in refining capacity and oil storage capacity in order to maintain its market position as a refining and oil-trading leader[170].

Singapore is the undisputed oil hub in Asia and along with the U.S. Gulf Coast and Rotterdam is one of the world's top three export refining centers[171]. Singapore's refining capacity is currently around 1.3MMBPD, with several plans for expansion. Singapore's storage capacity is around 129MMBBLs for both crude, refined products and other hydrocarbons[172].

Singapore's energy and chemicals industry ranks among the top 10 globally, and the industry employs more than 25,000[173]. Singapore is also the world's busiest marine bunkering center.

U.S. Refining Capacities

The Petroleum Administration for Defense Districts (PADDs) are geographic aggregations of the 50 States and the District of Columbia into five districts. During World War II, the Petroleum Administration for War, established by an Executive Order in 1942, used these five districts to ration gasoline. Although the Administration was abolished after the war in 1946, Congress passed the Defense Production Act of 1950, which created the

---

[168] Ibid, page 19

[169] https://www.eia.gov/beta/international/analysis.cfm?iso=SGP

[170] Ibid

[171] https://www.edb.gov.sg/content/edb/en/industries/industries/energy.html

[172] https://www.iesingapore.gov.sg/Media-Centre/News/2015/12/Oil-trading-growing-in-region--Vopak

[173] https://www.edb.gov.sg/content/dam/edb/en/industries/Chemical%20Engineering/energy-and-chemicals-yearinreview2016.pdf

Petroleum Administration for Defense and used the same five districts, only now called the Petroleum Administration for Defense Districts[174].

The United States has the largest refining capacity in the world and it is divided in terms of PADDs by the Energy Information Agency (EIA)[175] (capacities shown in MBPD as of June 2016):

Category	PADD I	PADD II	PADD III	PADD IV	PADD V	Total US
States Covered	DE, NJ, PA, WV	IL, IN, KS, KC, MI, MN, ND, OH, OK, TN, WI	AL, AR, LA, MS, NM, TX	CO, MT, UT, WY	AK, CA, HI, NV, WA	
Number of Refineries	9	27	57	17	31	**141**
Atmospheric Distillation Capacity	1,277	3,922	9,515	679	2,924	**18,317**
Vacuum Distillation	5,864	1,775	4,830	256	1,626	**9,073**
Thermal Cracking	82	577	1,636	90	599	**2,983**
Catalytic Cracking	499	1,339	3,118	210	887	**6,052**
Catalytic Hydrocracking	45	322	1,309	55	586	**2,318**
Hydrotreating & Desulfurization	1,030	3,867	9,319	592	2,675	**17,483**

Additional facts about refineries in the United States:

- The vast majority of refineries (close to 82%) in the US have a catalytic cracking unit. An FCC unit produces a higher yield of gasoline vs. distillate fuels since the local market demands *significantly* more gasoline than diesel. This is in contrast to markets like Europe, where diesel is in higher demand than gasoline[176].

- Less than 40% of refineries[177] in the US have any hydrocracking units, which makes sense since the US market demands *less* gasoline than diesel

- A large number of U.S. refineries have hydrotreating or desulfurization capacity, in particular to remove sulfur and other contaminants in order to meet ever rising product specifications[178].

---

[174] https://www.eia.gov/todayinenergy/detail.php?id=4890
[175] Refining Capacity Report, Energy Information Agency, June 2016
[176] Ibid
[177] Ibid

Refining Key Metrics Overview

Refining metrics are widely used by analysts, investors and companies to evaluate the *operational* and *financial* performance of a refinery or refining assets. In this chapter, the following refining metrics are introduced:

- Total crude oil processed
- Total throughput volumes
- Refined product yields
- Total refining capacity
- Refinery utilization
- Clean product yield
- Refinery complexity index
- Earnings per barrel
- Cash per barrel
- Market crack spread
- Realized crack spread

Total Crude Oil Processed

Total crude oil processed, also known as crude inputs, is the total volume of *crude oil* a refinery has processed for a specific period of time. Total crude oil processed can be measured in barrels or metric tons and is usually expressed in terms of either a daily or yearly quantity. Total crude processed volumes are usually compared against a refinery's crude distillation capacity to arrive at the *utilization* rate. It is important to note that total crude processed is *different* than total processed inputs, which is explained in the next section.

> *Refining Company ABC has three refineries in the United States. During 2016, refinery A processed a total of 30MBPD of crude oil, while refinery B processed 20MBPD of crude oil while refinery C processed a total of 100MPD of crude oil. Therefore, in 2016, Refining Company ABC processed 150MBPD of crude oil in total.*

Total Throughput Volumes

Total throughput volumes, also known as *total refinery inputs* or *total processed inputs*, is an overall measure of the volumes of hydrocarbons a refining company is processing, which includes crude oil *and* other feedstocks such as NGLs. The higher the total throughput volumes, the more refineries or

---

bigger refineries a company has. This metric is widely used to rank companies in terms of size.

> *In the year 2016, Refining Company ABC had total throughput volumes of 950MBPD of crude oil and 50MBPD of NGLs. Therefore Company ABC's total throughput volumes for the period were 1,000MBPD.*

Refined Product Yields

Total refined product yields is a volume-based metric, similar to total throughput volumes, but it measures the *actual production* of refined products that are produced at a refinery or at a company. The difference between total throughput and product yields is what is called refinery gain.

> *In 2016, Refining Company ABC's refineries produced 400MBPD of gasoline, 200MBPD of diesel, 50MBPD of jet fuel and 100MBPD of other products, including propane, asphalt, heavy fuel oil and residual fuel oil. Therefore, in 2016 total refined product yields for ABC were 750MBPD.*

Total Refining Capacity

Global Refining Capacity measures the *actual design* capacity of the refinery in terms of *total crude distillation capacity* of a refinery or refining company. It is the number one way to rank the *relative* size of a refinery or refining company. Global Refining Capacity is different from total throughput volumes since it is based on the *crude oil distillation capacity* instead of actual volumes being processed. Refining capacity, as it name implies, is the actual *design capacity* of the company's refineries. Total refining capacity can be compared with total throughput volumes to arrive at *how much* of the crude oil distillation capacity is being used currently (utilization rate, described on the next page). In times of high margins for refined products, a refinery is always looking to have as high a utilization rate as possible. Because of the seasonality of refining margins, particularly in the U.S. with the summer driving season, most refineries schedule their turnaround projects to be completed around the Autumn and Spring seasons since margins are lower, thus having the refinery incur a *lower opportunity cost* when completing a turnaround.

> *Refining Company ABC had 250MBPD of crude distillation capacity in Europe, 350MBPD in the U.S. and 100MBPD in Asia. Therefore the company's global refining capacity is 700MBPD.*

Several companies also calculate refining capacity differently than the standard, which is based on the capacity of *only* the crude distillation units.

An example is Valero, which calculates their refining capacity based on the capacity of the distillation units *and* other units. From their recent 10-K filing:

> *"Throughput capacity of 3MMBPD, represents estimated capacity for processing crude oil, inter-mediates, and other feedstocks. Total estimated crude oil capacity is approximately 2.5 million BPD."*[179]

Refinery Utilization

Refinery utilization rate is a measure that basically answers the question *how much* of the existing refining capacity is being used? The higher the utilization rate, the more volumes are being run through the existing refining capacity. A low utilization rate would indicate idled equipment, challenging economic conditions or simply reliability issues causing the refinery to not be able to process *as much* feedstocks as possible.

A low utilization rate might be due to several reasons:

- Refineries have had turnarounds during the period. Turnarounds are extensive maintenance programs that usually may shut down the entire refinery or certain production units of the refineries, affecting the utilization rate.

- Temporary or permanent shut downs[180] due to market conditions. If a refinery is not economical for a long period of time, the company may decide it is better to *temporarily* shut down the refinery for a period of time until market conditions improve. Another reason for a temporary shutdown could be due to a pending sale of the refinery itself[181].

- Feedstocks or crude oil input constraints, whereby a refinery cannot run at capacity due to feedstocks being mismatched. For example if a refinery was built to run more efficiently using heavy crude oils, but only light crude oils are available, this may cause bottlenecks in refinery processing units, such as distillation units, that may cause the refinery to not fully utilize all the installed capacity.

---

[179] Valero 2017 Form 10-K, page 2
[180] A good example of a recent shutdown was the Hovensa refinery in the U.S. Virgin Islands. Please see here: http://viconsortium.com/featured/hovensa-says-it-will-completely-shut-down-refinery-in-mid-december-if-abr-vi-deal-with-govt-fails/
[181] http://www.reuters.com/article/2011/09/27/us-conocophillips-trainer-idUSTRE78Q5R320110927

- Unplanned maintenance events, such as process upsets, equipment malfunction, operator errors, safety incidents and other unforeseen incidents that impact the utilization rate of refineries.

Refinery utilization is calculated by *summing* all refining inputs[182] or total throughput volumes and *dividing* those volumes by total refining capacity in terms of the atmospheric crude distillation units[183].

> *Refinery XYZ company had total refinery inputs of 300MBPD while its global refining capacity is 350MBPD. Therefore the company achieved a refinery utilization of 86%. The cause for the lower utilization versus a quarter ago was the fact that one of its refineries had an extended turnaround for two months.*

Refinery utilization is widely followed by investors as well as by research agencies such as the Energy Information Agency[184].

Clean Product Yield

Clean product yield consists of *adding* up total refinery production volumes of gasoline, jet fuel and diesel products and *dividing* these volumes by total throughput volumes.

Gasoline, jet fuel and diesel (also known as "clean products") are traditionally the most *profitable* refined products coming out of a refinery, especially in comparison with lesser value products such as heavy fuel oil (HFO), petroleum coke or asphalt. Clean product yield measures how the company's refinery configuration is able to produce more of these *higher valued* products, and thus be more profitable than competitors. A refinery that is equipped to increase the output of these *higher value* products, the higher the clean product yield will be and thus the higher earnings a refinery will have. Inversely, the lower the clean product yield, the overall *less* profitable the refinery will be. Clean product yield is also a measure of the complexity of the refinery. All things equal, the more complex the refinery is, the higher the clean products yield percentage will be.

---

[182] Total input to atmospheric crude oil distillation units. Includes all crude oil, lease condensate, natural gas plant liquids, unfinished oils, liquefied refinery gases, slop oils, and other liquid hydrocarbons produced from tar sands, gilsonite, and oil shale.
[183] Represents the utilization of the atmospheric crude oil distillation units. The rate is calculated by dividing the gross input to these units by the operable calendar day refining capacity of the units. Source: https://www.eia.gov/dnav/pet/TblDefs/pet_pnp_unc_tbldef2.asp
[184] http://www.eia.gov/dnav/pet/pet_pnp_unc_dcu_nus_m.htm

Clean product yield is calculated by *adding up* gasoline, jet fuel and diesel volumes and *dividing up* the results by total throughput volumes or total processed inputs.

> *Refining Company ABC produced gasoline volumes of 100MBPD, diesel volumes of 40MBPD and jet fuel volumes of 30MBPD, for total clean product volumes of 170MBPD. ABC reported total throughput volumes of 200MBPD. Therefore, ABC's clean product yield for this period was 170 divided by 200 or 85% during the period.*

Refinery Complexity Index

The refinery complexity index, also known as the Nelson Complexity Factor, is a factor to analyze the *relative size* and investment cost associated with a refinery. W.L. Nelson developed this concept in the 1960's in a series of articles for the Oil & Gas Journal[185], and measures the *relative costs* of processing units that make up a refinery in comparison to the most basic unit of a refinery, the atmospheric distillation unit.

The higher the nelson complexity index, the more complex the refinery is and the more heavier crude oils a refinery can process as well the higher upgrading capacity to process heavier hydrocarbon molecules into *higher valued* transportation fuels.

The following table provides the factors assigned to the various processing units[186]:

Unit	Complexity Index Factor		Unit	Complexity Index Factor
Atmospheric Distillation	1.0		Catalytic Cracker	6.0
Vacuum Distillation	2.0		Catalytic Reformer	5.0
Visbreaking	2.5		Catalytic Hydrocracking	6.0
Thermal Cracking	3.0		Alkylation / Polymerization	10.00
Thermal Processes	5.0		Aromatics / Isomerization	15.00
Delayed Coking	6.0		Lubricants	60.00

---

[185] http://www.ogj.com/articles/print/volume-94/issue-12/in-this-issue/general-interest/refining-report-complexity-index-indicates-refinery-capability-value.html
[186] http://pakpas.org/0.REFINERY%20LIBRARY/2.EDC-business_petroleum_refiningmktg_lc_ncf.pdf

The index is calculated by starting with the atmospheric distillation unit which is always assigned an index of one. All the other units are calculated in a *weighted average basis* as a percentage of the distillation capacity. The following table provides an example on how to calculate the complexity for a large size refinery:

Unit	Capacity (MBPD)	% of Crude Capacity	Complexity Factor	Calculated Complexity Index
Atmospheric Distillation	600	100	1.0	1.00
Vacuum Distillation	300	50	2.0	1.00
Coking	60	10	5.5	0.55
Catalytic Cracking	144	24	6.0	1.44
Catalytic Reforming	102	17	5.0	0.85
Catalytic hydrocracking	30	5	6.0	0.30
Refinery Complexity Index				**5.14**

Earnings per barrel

Earnings per barrel is a financial measure of how *profitable* a refining company is on a GAAP *accrual accounting basis*. A company with a high quality portfolio of complex refineries (i.e. with a high nelson complexity index, high clean product yield, access to discounted feedstocks) that runs efficiently controls costs and increases revenues would tend to have higher earnings per barrel than its competitors. Please note that earnings may have to be adjusted to account for "Special items", which are items such as asset gains or asset impairments that do not reflect *"core earnings"* but are still accounted under GAAP rules as *earnings*.

The adjusted earnings per barrel metric is calculated two ways. If a company or segment has embedded marketing operations, then total refined product sales are used as the denominator:

- GAAP earnings (adjusted for non-cash or special items) for the period *divided* by total refined product sales.

  Refining company ABC had earnings of $200MM, which included an asset impairment of $40MM and had total refined product sales of 10 million barrels for the period. Therefore, adjusted earnings were $240MM, which is then divided by total refined product sales for the period of 10 million barrels, which equals to adjusted earnings of $24 per barrel.

In cases where a refining company doesn't own a marketing business or does not disclose refined product sales, then total throughput volumes can be used instead as the denominator.

- GAAP earnings (adjusted for non-cash or special items) for the period *divided* by *total throughput* volumes.

    Refining company ABC had earnings of $100MM, which included an asset impairment of $20MM and had total throughput volumes of 6 million barrels for the period. Therefore, adjusted earnings were $120MM, which is then divided by total throughput volumes for the period of 6 million barrels, which equals to adjusted earnings of $20 per barrel.

Throughput this book, total refined product sales is *predominantly* used as the denominator, although in certain instances, total throughput volumes are used instead. The reason for this is that most large companies tend to combine their refining & marketing operations into one segment for external reporting purposes.

Cash per barrel

Cash per barrel is a metric that measures of *how much* cash is being generated by a refining company on a per *barrel* basis. Total cash (the numerator) is calculated as follows:

- Downstream or Refining segment GAAP earnings *plus*
- Adjusting or special items *plus*
- Depreciation & Amortization (D&A)

Similar to earnings per barrel, the denominator for this metric depends on whether a refining company has marketing operations or not. If a refining company does have marketing operations, the denominator will be total refined product sales. If a refining company does not have marketing operations, total cash will then be divided by *total throughput volumes*.

Refining Company ABC only has refinery operations and does not have wholesale marketing operations. In 2016 ABC had earnings of $200MM, no special items and D&A of $50MM while total throughput volumes were 10MM barrels. Cash per barrel is therefore $25.

Integrated downstream company XYZ has both refinery and marketing operations. Last year, they had $300MM of earnings, D&A of $150MM, while refined product sales were 15MM barrels. Cash per barrel for the period is thus $30.

The reason cash per barrel is calculated by adding back depreciation & amortization to earnings *instead* of using *actual cash flow from operations*, is that cash flow from operations includes working capital fluctuations which add unwanted fluctuation and noise to this metric.

Market Crack Spread

The market crack spread is one of the quickest and easiest ways to understand *overall* refining profitability at a particular point in time. The reason this metric is called the "crack spread" is that a refinery basically *chemically cracks* or transforms *long* chain hydrocarbons into *smaller*, more valuable *shorter* chain hydrocarbons that are more useful for end products like gasoline or diesel. For example, gasoline can be said to be comprised of hydrocarbons primarily ranging from 5 carbon molecules (C5 or pentanes) to 9-10 carbon molecules (C9-C10), while a typical crude oil might be composed primarily of C12-C20s or so. The job of a refinery then is to most efficiently *transform* raw crude oil into finished products like gasoline, diesel, jet fuel and other products. In other words, this metric captures the market *differential* between the prices of refined products vs. the price of crude oil.

The crack spread simply reflects the difference between the purchase price of crude oil and the price of refined products. There are several variations, but the most commonly used is the *3-2-1 crack spread*. This crack spread assumes that a typical refinery running *3* barrels of crude oil will yield *2* barrels of motor gasoline and *1* barrel of diesel. Refining crack spreads change *every time* prices change for crude oil or for refined products. With *futures* market like the Chicago Mercantile Exchange (CME) or New York Mercantile Exchange (NYMEX), futures contracts can be used to analyze the market assessments of future crack spreads[187].

Refining crack spreads can vary widely depending on which benchmark crude oils and refined product sales prices are used. An additional variable that impacts the crack spread is *which* locations are used to calculate these prices.

To calculate an example *3-2-1* crack spread, the following commodities are used:

- West Texas Intermediate (WTI) crude oil
- NYMEX New York Harbor gasoline (NYH)
- NYMEX Ultra-Low Sulfur Diesel (ULSD)

---

[187] For more information, please visit the CME's crack spreads handbook:
http://www.cmegroup.com/trading/energy/crack-spread-handbook.html

In the month of July, WTI crude oil was $50/bbl, NYMEX gasoline was $2 per gallon ($84/bbl), and NYMEX ULSD was $1.8 per gallon ($75.6/bbl). Using the 3-2-1 we arrive at the following calculation:

Refined product sales: 2 barrels of gasoline times $84 equals $168, 1 barrel of diesel at $75.6, for a total refined product sales of $243.6 Minus total WTI crude oil costs of 3 barrels times $50 or $150 Thus the market crack spread is then $93.6 divided by 3 or $31.2 per barrel.

Realized Crack spread

Many companies provide this metric or can be easily calculated, which compares *how much* of the average market crack spread a company is "capturing" or realizing.

For refining companies or for refining segments of oil & gas companies, the realized crack spread can be calculated *as long as* an income statement is provided for the refining company or refining operations and crude volumes processed, total throughput volumes or refined products production are reported. The realized crack spread is typically calculated as follows:

- Refining Sales or Revenues *minus*
- Refining Cost of Goods Sold, the result which is then *divided by*
- Total refined products production or total throughput volumes in barrels

  Refining company ABC had Refined Product sales of $10 billion in 2016, while cost of goods sold for this company was $9 billion in the same year. Overall refined products production was 400MBPD or 146 million barrels per year. $10 billion minus $9 billion equals gross margin of $1 billion divided by 146 million barrels equals $6.85, thus ABC's realized crack spread was $6.85 per bbl in 2016.

Many of the differences between realized and market crack spreads arise due to:

- Differences in refining configuration, not all refineries have the typical "3-2-1" configuration and could have different product yields of refined products. Another reason could be that the company's refineries, based on the current installed equipment at the refineries, may be producing more of the lower valued products such as asphalt or pet coke.

- Crude differentials between a benchmark crude oil like WTI or Brent to the actual crude slate that the company's refineries are using. For example a refinery might be using more expensive crude oils like LLS vs. WTI because it is better suited for the refinery's existing equipment configuration.

- The fact that the market crack spread does not account for the production of *secondary products* such as natural gas liquids, fuel oils or petroleum coke. In other words, the 3-2-1 market spread assumes that the only two products being produced out of a refinery are gasoline and diesel/jet fuel, which is not what occurs in real life operations.

- Combination of both crude slate and refinery configuration: For example if a refinery is using more of the *heavier, cheaper* crude oils and the refinery is yielding more diesel than gasoline this will affect both the *feedstock* side of the crack spread equation as well as the *refined products* side of this equation.

- Refinery locations: If a company has refineries located in *cost advantaged* locations from a crude feedstock perspective (i.e. close to a stranded source of crude oil with limited takeaway capacity) or from a refined products supply center (located close to an area with high demand for refined products), the company's realized crack spread will be substantially *higher* than the market crack spread. The same would apply if the opposite circumstances exist (i.e. refineries located in disadvantaged locations would have crack spreads *lower* than the market crack spread).

- Effectiveness of the company's supply & trading organization: How effective the company's supply & trading organization is in buying *cheaper* crude oils and selling *higher value* refined products vs. what the average market participant is buying and selling these products for.

The realized crack spread is also provided by several companies and can be also stated as a percent of a benchmark crack spread:

> *In the second quarter, the market crack spread was $25/bbl, while Refining Company ABC's overall realized crack spread was $30/bbl or 120%. The higher realized crack spread over the market was primarily attributable to several refineries' configuration that produced more higher-valued diesel during the quarter.*

Cash Flow Profile of a Refining Asset

A typical cash flow profile for a refining asset can be summarized into the following steps:

- Conduct economic and feasibility studies (cash outflow).
- Refining site is acquired (cash outflow).
- Refinery construction starts (cash outflow).
- Refinery construction ends (cash outflow).
- Purchase of feedstock inventory *prior* to the refinery being operational (cash outflow).
- Payment of on-going feedstocks, operating expenses and sale of refined products (cash inflow from sales of products *minus* feedstock costs *minus* operating expenses).
- Turnarounds or extended maintenance programs are conducted every three to seven years (cash outflow).
- In times of good margin environments, revenues consistently exceed cost of goods sold and operating expenses.
- In times of low economic growth, revenues might only cover cost of goods sold and operating expenses.
- In times of economic recessions, revenues might only cover cost of goods sold but not enough to pay for operating expenses, causing some refineries to shutdown either temporarily or permanently.

Cash Flow profile for a Refining Asset
Low Complexity Refinery

Year	$MM (Outflow)/Inflow	Description
2010	(750)	Refinery Construction Begins
2011	(500)	Refinery Construction Continues
2012	(250)	Refinery Construction ends year-end
2013	400	Refinery Operational
2014	600	Margins Improve
2015	150	Economic Recession, Margins Decline
2016	(100)	Continued Low Margin Environment
2017	(150)	Low Margins worsen, refinery closures
2018	200	Low Crude Costs improve margins
2019	400	Gasoline and Diesel prices improve
2020	500	Margins Improve
NPV @10%	$21.60	

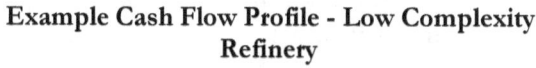

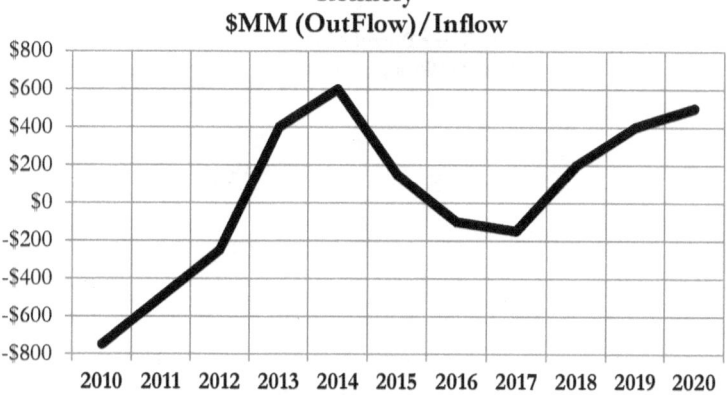

Business Cycle in Refining

The refining business, along with upstream, is the *most* cyclical of all the sectors in the oil & gas industry. As can be seen from the chart below[188], refining margins fluctuate heavily every day and have recurrent up and down cycles:

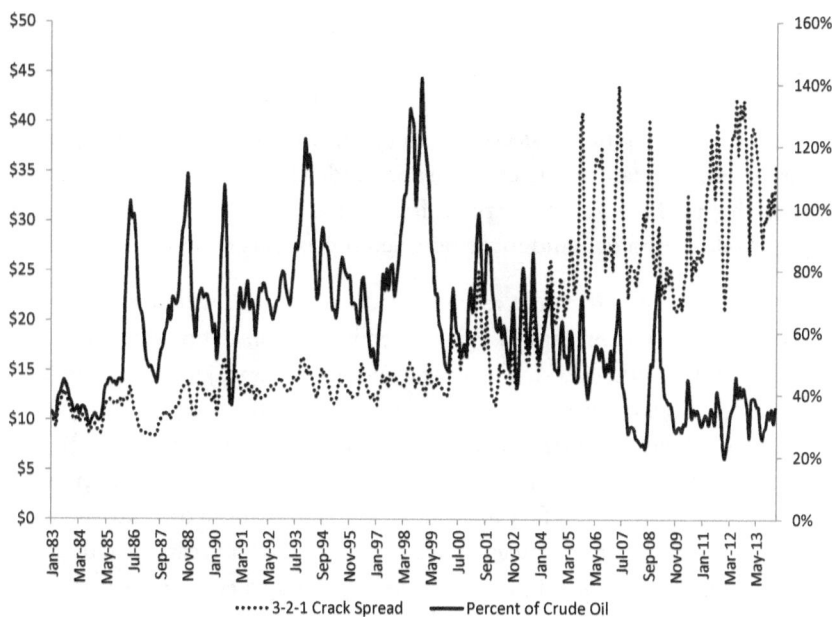

[188] Created using data from the Energy Information Agency's website: http://www.eia.gov/petroleum/data.cfm#prices

The second chart below shows how much, in 2009 dollars per barrel, the average margin for refining companies has fluctuated since 1977:

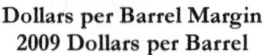

Dollars per Barrel Margin
2009 Dollars per Barrel

As an example, in the years before the 2008 economic crisis, refining margins were above $4/bbl. After the 2008-2009 economic recession refining margins decreased substantially and became negative for some part of 2009. Refining business cycles can last from 3 to 6 years and depend heavily on the state of the economy at hand. The more prosperous the economy is, the *higher* the demand for refined petroleum products, as people drive and demand more petroleum products overall.

In the long term, growth in refined products demand will increasingly come more and more from rapidly growing economies in Latin America, Africa, Asia and the Middle East. The economies of Europe, Japan and the U.S. will, over time, reduce their absolute demand for refined products due to ongoing efficiencies (lower energy consumption per dollar of GDP) as well as economies that are not growing as fast as those in the developing world. We can all relate to the fact that personal vehicles these days are significantly more fuel efficient than vehicles 20 or 30 years ago, therefore, the demand trend for refined products in advanced economies is generally *trending* lower over the next several decades.

Why invest in Refining?

The refining side of the business is critical to the oil & gas industry in general as well as to the world economy. Without usable refined products being sold around the world, crude oil & natural gas in their *raw unprocessed* forms would not have valuable uses.

Despite refining being a cyclical and margin business, many successful companies with a large downstream presence, like ExxonMobil, have managed to combine the large cash generation ability of downstream with the high returns *historically* inherent in E&P operations.

The resilience of the integrated model is best summarized by the following quote from Rex W. Tillerson, former Chairman and CEO of ExxonMobil:

> *"ExxonMobil's quarterly results demonstrate the strength of our integrated business model. Integration across Upstream, Downstream and Chemical gives us competitive advantages in scale, efficiency, technical and commercial capabilities, regardless of market fluctuations over the business cycle"*[189]

One big advantage that refining, from a financial perspective, has over upstream, is the fact that once a refinery, terminal or pipeline is built, there is a *long cycle* between when additional capital expenditures are needed to keep existing refineries in operating condition, which is quite different from E&P. For a simple refinery, a turnaround cycle can usually be from 3 to 7 years or so; in comparison to the total cost of the refinery, turnarounds or additional capital expenditures are *relatively low*. By comparison, in the E&P business high *on-going* capital expenditures are required every year *just* to be able to maintain *existing* levels of production.

Another good fact to mention is when looking at an integrated oil & gas company, the downstream side of the business will usually have the highest amount of revenues in comparison to upstream. This is due to the fact that sales to third parties usually occur from a downstream perspective. In other words, if the E&P segment sells crude oil to the refining segment of an integrated company, the ultimate sale will be recorded as refined product sales of say gasoline and the crude sale from E&P to refining will be eliminated in the *consolidated* financial statements[190]. An additional reason is the historical fact that downstream is traditionally the sector of the oil & gas

---

[189] http://news.exxonmobil.com/press-release/exxon-mobil-corporation-announces-estimated-third-quarter-2014-results
[190] See example: Total 2013 20-F Form, page F-9 "Principles of Consolidation"

industry that tends to have the highest amount of revenues, but has significantly lower margins than E&P operations.

As every day consumers of petroleum products, we are beneficiaries of having downstream companies who supply each service station with fuel products; that is products that can be used for personal transportation and products that are critical to the transportation of goods and services in the global economy.

As investors, refining companies present unique opportunities, such as having large amounts of operating revenues and generating substantial cash flows. A refining company can therefore play a key role in an investor's diversified portfolio.

Chapter IV – Petroleum Marketing & Trading

"To think creatively, we must be able to look afresh at what we normally take for granted."– George Kneller

Marketing & Retail Overview

The petroleum marketing is usually the most *customer-facing* of all the businesses in the Oil & Gas industry.

Downstream companies usually have marketing operations in order to guarantee *product outflow* from the refineries so that these refineries can run as reliably as possible and not encounter any product bottlenecks down the road that could impact the refinery's operations.

Marketing operations can purchase petroleum products either from their own company's refining operations or from third parties. These purchased petroleum products are then resold and marketed at company-owned, franchised or even third party retail outlets. Many large integrated companies, such as RoyalDutchShell or ExxonMobil have a large global retail presence[191]. There are also smaller fuels marketing companies that may have local, regional or even national marketing operations.

History of Petroleum Marketing

The history of petroleum marketing is as old as *exploring & producing* oil & gas, beginning with the first commercially successful oil well in 1859 in Pennsylvania. Since the invention of the modern combustion engine by Karl Benz in the late 19th century and the later mass production of vehicles with the Ford Model T, global gasoline demand experienced a very fast growth in the beginning of the 20th century. The mass production of automobiles led to an increase of demand for building new service stations across the world.

From commercializing the first barrel of oil, to marketing lamp oil (kerosene) for lighting, to marketing transportation fuels the value proposition for this business has always been:

- Buying the *lowest cost* inputs into a refinery processes or into the company's fuel marketing operations.
- Providing access to consumers to purchase fuels and products at a price that is *competitively* priced to clear the market.
- Ensure supply of crude oil, refined products and feedstocks to whichever areas have highest value for these.
- Deliver these commodities to the areas where end customers, marketing or trading companies require these products.

---

[191] Please note that these companies do not necessarily own or operate these gas stations.

Motor Vehicle Ownership & Fuel Demand

Motor vehicle ownership has grown substantially since the beginning of the 20th century, creating significant demand for transportation fuels. In the United States, at the beginning of the 20th century, there were around 8,000 registered vehicles in the entire country, increasing to more than 250 million vehicles in 2014[192]:

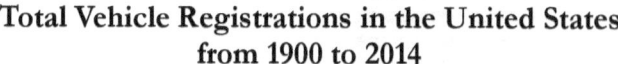

Total Vehicle Registrations in the United States from 1900 to 2014

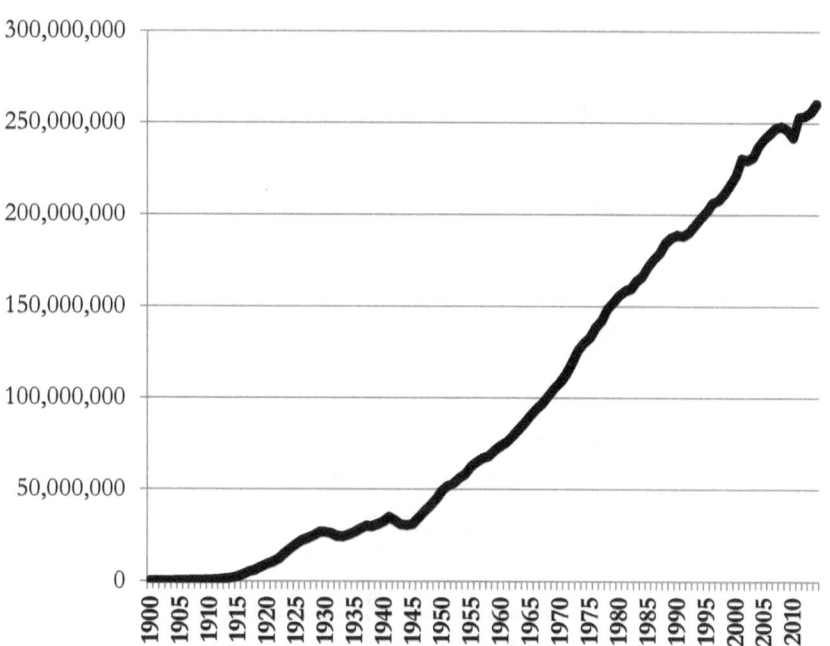

In the United States, the average American drives 32 miles a day, using about 22.27 barrels of oil per year, which is reasonable given that 85% of Americans go to work by driving an automobile[193].

After many years of rapid growth, China has surpassed the United States in new total vehicle sales and in 2016 close to 28 million vehicles were sold in China[194], this is in comparison with the United States with vehicle sales of 17.5 million for the same year.

---

[192] Source: U.S. Department of Transportation, Federal Highway Administration, Vehicle Registrations. https://www.fhwa.dot.gov/policyinformation/quickfinddata/qfvehicles.cfm
[193] National Association of Convenience Stores, 2015 report, pages 5 & 6
[194] https://www.bloomberg.com/gadfly/articles/2017-03-20/china-s-car-sales-are-running-out-of-gas

Both the United States and China have more than 200 million registered vehicles in their countries.

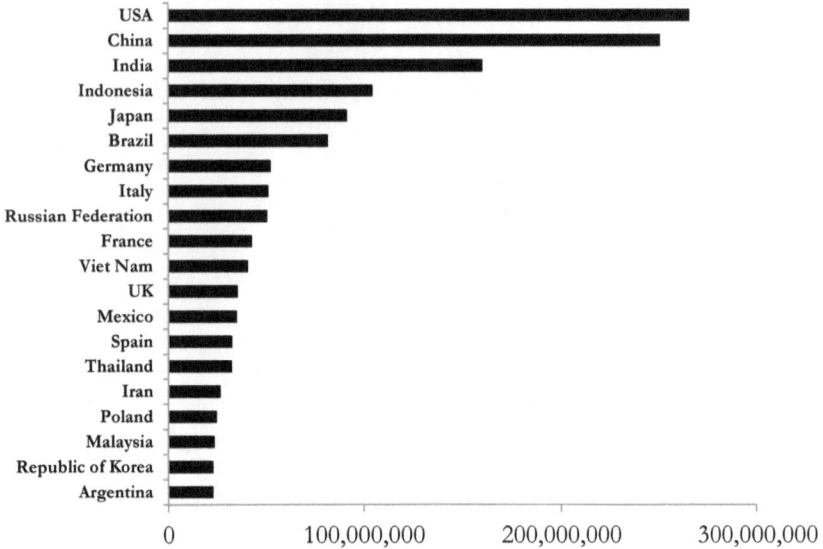

Top 20 Countries with Most Vehicles
Source: World Health Organization

Although China is now the largest vehicle market in the world, China ranks number 92 out of the countries with most vehicles per 1,000 people[195]:

Rank	Country	Vehicles per 1,000 people	Rank	Country	Vehicles per 1,000 people
1	San Marino	1736	7	Luxembourg	813
2	Monaco	1085	8	Malaysia	802
3	Finland	1080	9	Malta	753
4	Andorra	964	10	Austria	752
5	Italy	841	92	China	181
6	USA	828	108	India	127

The low ownership per capita of automobiles in markets like China and India presents opportunities of higher fuel consumption by these countries, which will continue to impact demand for crude oil around the world for many years to come.

---

[195] Source: World Health Organization, number of register vehicles
http://www.who.int/gho/road_safety/registered_vehicles/number/en/

Businesses within Marketing & Trading

- Wholesale marketing, which is involved in *buying* refined products produced at refineries and *selling* these products on a wholesale basis through a rack marketing network. Typical customers of wholesale marketing are wholesalers, distributors, retailers (service station operators) and truck-delivered end users.

- Retail marketing, which is involved in retail sales of petroleum products as well as operating, leasing or owning convenience stores to sell merchandise as well as fuel.

- Supply, which is involved in purchasing crude oils, feedstocks and petroleum products at the *lowest cost* while selling petroleum products and feedstocks at the *highest value* and supplying these petroleum products at the time and location that they are needed.

- Trading, which is involved in *optimizing* inventories by *buying* and *selling* crude oils, feedstocks and products that are not needed for the base operations of a refining & marketing company.

Wholesale Marketing

Wholesale marketing involves offering both *branded* and *unbranded* refined products on a *wholesale* basis through an extensive rack marketing network. The principal customers of a wholesale marketing organization are large distributors, retailers and truck delivery companies, typically through long-term contracts.

Usually, a significant majority of the fuel sold by refiners is sold through wholesale marketing to large customers and distributors such as:

- Airline and marine customers.
- City, State and Federal Government agencies.
- Customers with large commercial fleets, such as local bus service companies, trucking companies, railroad companies and many others.
- Resellers or distributors, which pick up fuel at the terminal rack and distribute fuel to end retailers that may be branded or unbranded customers.
- Utilities.

Retail Marketing

Retail marketing, particularly in the United States, is highly competitive, fragmented and where there are no significant large single players. Most of the integrated oil & gas companies have exited the retail marketing business and new independent companies, like Kwik Trip, QT, Buckee's as well as grocery stores across the United States, have entered the retail market and compete with the traditional branded retail stations where drivers buy fuel.

Number of Service Stations for Select Countries

The following table provides a sample of selected countries and the number of service stations:

Country/Area	Number of Service Stations	Source
United States	152,995	National Association of Convenience Stores
European Union	114,431	Fuels Europe.eu
China	99,000	Various sources
Brazil	40,880	Petrobras
Japan	34,000	Bloomberg
Canada	11,916	Canadian Fuels
Mexico	11,438	Comisión Reguladora de Energía
Russia	12,000	Estimated

Typical Revenue & Expenses items in the service station business

The following are typical *revenue* and *expense* streams in the service station business[196]:

Area	Item	
Revenues	• Fuel Sales	• Lottery
	• Convenience Store Products	• Car Wash
	• Food Service	• Money Orders
	• ATM Fees	• Rent Income
	• Air Machine	• Vendor Rebates
Expenses	• Advertising and Promotions	• Merchandise (Over/Short)
	• Bad Debts	• Miscellaneous
	• Cash (Over/Short)	• Operating Supplies
	• Commissions	• Overhead (General and
	• Credit Card Fees	Administrative)
	• Debt Service (Principle Payments)	• Payroll Tax
		• Pest Control
	• Environmental	• Professional Fees
	• Fuel (Over/Short)	• Property Tax
	• Hiring and Training	• Rent
	• Insurance	• Repairs and Maintenance
	• Interest	• Robbery and Theft
	• Leases	• Store Benefits
	• Legal Fees	• Telephone
	• Licenses	• Utilities

---

[196] http://www.phillips66fuelsupplier.com/improve-your-business/sales-potential/

Branded

The branded marketing business one of the most customer facing businesses in the oil & gas industry and that intrinsically has been associated with the oil & gas industry at large. Most oil companies no longer own or operate retail fuel stations but instead provide *branding agreements* to independent owners who sign contracts with major brands. However, still close to 50% of the fuel sold in the United States is still branded[197]. Branded marketing groups typically provide a downstream organization with a *price uplift* since branded fuel provides a selling price *premium* over unbranded fuels.

Additives

Branded fuels typically have company proprietary additives which exceed minimum requirements. Additive brands such as *Techron*, *V-Power* or *Synergy*, command a higher price over unbranded fuels with minimum additives or other non-premium brands. Additives typically provide protection against:

- Reduction of *gunk* or carbon deposits that overall make combustion less efficient.
- Reduction of corrosion of metal parts that are caused by moisture and other contaminants. Fuel additives provide a *reduction* of the impacts of corrosion to a vehicle's engine.
- Reduction of wear or friction inside the combustion engine's cylinder walls, which can lead to higher oil consumption, loss of power and higher emissions[198].

Additives work via the addition of several components that typically fall under the following:

- Detergents, which help provide protection against intake valve deposits as well as deposits in the fuel injection system[199].
- Anti-adhesion compounds which help prevent detergents from forming a film on intake valves.
- Corrosion inhibitors, which inhibit corrosion in the fuel system.
- Demulsifiers, help prevents water being mixed with the fuel.
- Solvent fluids, enables the additive ingredients to stay mixed and to flow, and not freeze at very cold temperatures.

---

[197] NACS Online – Who Sells America's Fuel, published 03/01/2017
[198] http://www.shell.us/motorist/shell-fuels/shell-v-power-nitro-plus-premium-gasoline/shell-v-power-nitro-plus-premium-gasoline-faqs.html
[199] https://www.exxon.com/en/unleaded-gasoline

Another marketing dynamic that companies provide to increase the sale of higher octane gasolines is the addition of a *higher* number of additives to these premium higher-octane fuels.

Pricing

Pricing is usually sold from the marketing organization to stations or dealers at what is commonly called *dealer tank wagon* or DTW, which is higher than what are called *rack* prices. DTW volumes are typically delivered and custody transfers to the retail owner at the service station[200].

Company Owned & Operated

When service stations first started, this was the most common type of arrangement, where a large integrated oil & gas company would own, operate and supply fuel to a gas station. Interestingly enough, the vast majority of the public assumes that service stations are owned and operated by the brand used, but this is no longer the case, especially in the U.S[201].

Lessee/Lessor relationship

The lessee owns the *business* while the downstream company owns the *land* and *building*. The lessee pays rents to the downstream company while it operates the station while providing the company a *guaranteed* supply outlet. This type of arrangement assures the downstream company a *reliable* and *ratable* supply outlet for its petroleum products[202]. The lessee owns *all* motor fuel and convenience store items and retains all profits generated from the sale of fuel and items[203].

Dealer Owned & Operated

The dealer owns the *business*, *land* and *building* while the dealer purchases fuel from the downstream company. The downstream company usually has a marketing agreement whereby certain requirements are prescribed, such as number of pumps, signs, rebates and other conditions.

Commission or Consignment Sites

This is a more recent type of arrangement where the sites may or may not be owned by the downstream company but the key differentiator is the

---

[200] http://www.nacsonline.com/YourBusiness/FuelsCenter/Operations/Articles/Pages/How-Branded-Gasoline-Stations-Work.aspx

[201] For example, ConocoPhillips, back in 2008 when it was a still an integrated Oil & Gas company, sold all of its service stations http://www.chron.com/business/energy/article/ConocoPhillips-is-latest-to-sell-its-gas-stations-1643085.php

[202] http://www.nacsonline.com/YourBusiness/FuelsCenter/Operations/Articles/Pages/How-Branded-Gasoline-Stations-Work.aspx

[203] CST Brands Inc. 2016 Form 10-K, page 9

ownership of the *fuel inventory*. Since fuel prices are highly volatile and inventory fluctuations can generate big fluctuations in *working capital* requirements for independent owners, the inventory ownership and fluctuations impacts are absorbed by the downstream company while the owner can focus on growing fuel volumes and getting paid a *fee* per volume of fuel sold[204]. The site owner operates and retains all profits of all non-fuel related operations.

Contracts

Branded fuel contracts can vary significantly, but they share certain key elements[205]:

- Length: A typical contract is for 10 years, although contracts could be longer or shorter.
- Volume requirements: Contracts typically set forth a certain volume of fuel each month that retailers must buy.
- Image requirements: A branded retailer receives marketing support from the downstream company it signs the contract with, which may include advertising for the fuel brand. In addition the oil company may provide financial incentives to display its brands. Typically, the downstream company may also require certain imaging requirements and quality control.
- Wholesale price requirements: A branded retailer must purchase fuel from a branded supplier or distributor. Branded contracts benchmark the wholesale price to common fuels indexes, such as Platt's, plus a premium of a few cents for marketing support.

Unbranded

Unbranded service stations are owned or operated by companies such as 7-Eleven, QuikTrip, Wawa, as well as what are considered "big box" retail stores that include major grocery stores, as well as Wal-Mart, Sam's Club and others, which account for roughly 50% of fuel sales in the United States.

These companies purchase *unbranded* fuel from downstream companies or in the open market. These companies usually have *lower* fuel prices in exchange for the fuel not having as many additives as the typical branded fuel. It is important to note that since gasoline is a commodity, the very

---

[204] Ibid, page 9
[205] NAICS 2016 Retail Fuel Report, page 4

same gasoline molecules from an unbranded station to a branded station may be produced from the *very same* refinery. The key difference between unbranded and branded lies primarily in each company's branded proprietary added to the fuel, like *Techron* from Chevron, or *V-Power* from Shell as examples. The other difference is that typically, diesel fuel tends to be sold more on an *unbranded basis* from refining companies since diesel is seen as a commodity and is perceived purely as a *utility* fuel by those demanding it. This is particularly applicable in the United States, where the majority of demand for diesel fuel is driven by commercial and industrial users, such as long-haul trucking companies, railroads and other large scale users.

Unbranded volumes are usually sold at rack prices, or the prices at which wholesalers can buy fuel at the terminal rack, which is usually the disposition or transfer point between a downstream company and wholesalers or *unbranded* customers.

Comparison – Unbranded vs. Branded

There are several pros and cons for a retail fuel station to choose between unbranded and branded when opening up a new station[206]:

	Unbranded	Branded
Pros	• Lower Fuel price • No required maintenance or branding image requirements • No contract • No additional credit card fees • Flexibility on choosing fuel supplier • Fuel price negotiation	• Recognizable image and brand • Fuel additives • Trusted more by consumers
Cons	• Potentially unrecognizable image or brand • Station usually less attractive • Rare price inversion	• Expensive Price for fuel • Required to meet strict brand maintenance requirements • Tied to a contract with no way out • No fuel price negotiation • High credit card fees

The case for unbranded fuels, particularly in the United States, is that customers are very price sensitive and typically look for the *least expensive* fuel and *most convenient* location. Year after year, consumer surveys have shown that the number one factor when consumer buy fuels is price, with typically more than 70% of consumers choosing gasoline based on price

[206] http://desertfuels.com/industry-education/unbranded-vs-branded-gas-stations/

alone, with the next factor, for 20% of consumers, being the location of the store[207].

Marketing & Trading Supply Chain

The following table provides an overview of the petroleum marketing & trading supply chain, starting with a refinery and finishing up all the way to the service station[208]:

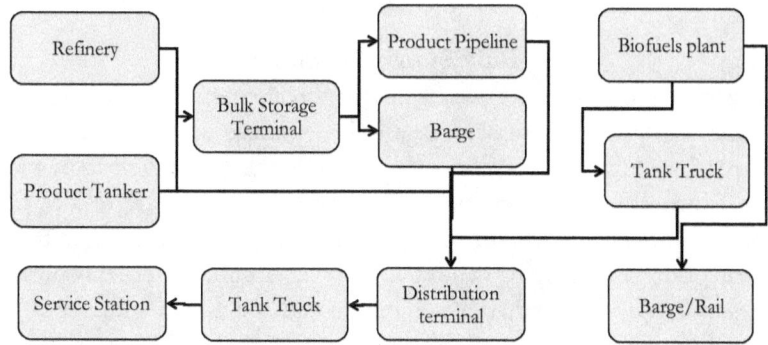

Product Pipelines

Product pipelines are the preferred and most *cost effective* method of transporting products inside many different countries around the world, but it is particularly used in the United States, where it accounts for 73% and 83% of all gasoline and diesel movements inside the country[209]. About 71% of all crude oil in the US is transported by pipeline.

Pipelines are also used to move crude oil from the wellhead to gathering and processing facilities and from there to refineries and tanker loading facilities. Product pipelines ship gasoline, jet fuel, and diesel fuel from the refinery to local distribution facilities. Pipelines require significantly less energy to operate and have the lowest incident rate in the industry when compared to other methods of transportation[210].

---

[207] http://www.nacsonline.com/YourBusiness/FuelsReports/2015/Documents/2015-NACS-Fuels-Report_full.pdf
[208] Assessment of the U.S. fuels distribution Network, from the Fuels Institute, page 19
[209] Ibid, page 23
[210] http://www.shipxpress.com/blog-article?r=b33gvcu3k8

Product Tanker and Barges

Barges and tankers, in the US, account for about 15-25% of all product movements.

A typical 30,000-barrel tank barge can carry the equivalent of 45 rail tank cars at about one-third the cost. Compared to a pipeline, barges are cheaper by 20-35%, depending on the route. And barge operators can transport crude oil down the Mississippi to Gulf Coast refineries and haul refined products back up the river to chemical plants and other end-users[211].

Bulk Storage Terminal

Bulk storage terminals are typically held at three types of location: refineries, pipelines and ports. Refinery storage holds gasoline and diesel fuel that has been produced but yet to be sent to market. Pipeline storage can be located at any point along the pipeline infrastructure, but typically resides at the point of product entry. Port storage is located at any U.S. port to store imports prior to moving to market. Once the gasoline or diesel is produced domestically or imported, the volumes could touch any number of bulk storage locations before moving to market[212]. Gasoline is usually stored as *blending* components in bulk storage terminals, not as finished motor gasoline since ethanol and other blending components have to be added at the final distribution point, which is the distribution terminal. One of the reasons for this is the fact that ethanol is a highly corrosive substance, due to the fact that ethanol is an alcohol. Alcohols are *hydrophilic* (like to attract water molecules), thus ethanol has to be trucked and cannot be transported by pipeline. Ethanol is corrosive and can degrade plastic or even metal parts and therefore cannot be blended in motor gasoline exceeding certain proportions as vehicle engines would have to be reengineered with alcohol resistant components[213].

Distribution Terminal

Distribution terminals are the primary access point between refining and marketing companies. These types of terminals are usually located near a pipeline, railroad, rivers or ports. Terminals generally receive refined petroleum products by pipeline, tanker or barge. The main function of these terminals is to provide temporary storage for refined products as well as to provide fuel dispending solutions for trucks distributing products to retail stations. These terminals usually transfer product to distribution

---

[211] http://www.shipxpress.com/blog-article?r=b33gvcu3k8
[212] Assessment of the U.S. fuels distribution Network, from the Fuels Institute, page 21
[213] http://www.popularmechanics.com/cars/hybrid-electric/a6244/e15-gasoline-damage-engine/

trucks at the *rack* point, which is why prices between a downstream company and a distributor are quoted at the "rack price". A terminal facility will usually store a variety of refined products: gasoline, heating oil, kerosene, diesel, jet fuel and others. Another key function that distribution terminals serve is provide blending capabilities for ethanol, since as previously mentioned, ethanol cannot be transported by pipeline, therefore it must be transported by truck or rail from an ethanol facility to a distribution terminal.

Trucks

Trucks are the most limited oil transportation method in terms of storage capacity, but they have the greatest flexibility in possible destinations. A typical truck can hold anywhere from 100 to 200 barrels of crude oil or petroleum products.

Trucks are often the last step in the transportation process, delivering oil and refined petroleum products to their intended storage destinations. These reliable and dependable means of transport are always working in the field, traveling between tank batteries, refineries, terminals and service stations. These trucks ensure petroleum products get delivered to a buying customer, provide market access to crude oil producers and royalty owners[214].

Biofuels Plant

Biofuels, such as ethanol or biodiesel, are produced to be later blended or sold as fuel in the United States and many countries around the world. In the U.S., since the passage of the Energy Policy Act of 2005 which required motor gasoline to be blended with a certain percentage of ethanol, ethanol has to be added to gasoline at about 10% of the volume[215]. Most vehicles on the road today can run up to 15% ethanol as per the manufacturers' specification, but blending a higher percentage than 15% would impact the engine, the performance of most vehicles and possibly void the car's warranty.

Renewable Identification Numbers

In the US, under the Renewable Fuel Standard Program, the EPA requires *refiners*, *marketers* and *importers* of petroleum products to use what is called *Renewable Identification Numbers* or RINs as credit for compliance under this RFS program.

---

[214] http://www.shipxpress.com/blog-article?r=b33gvcu3k8
[215] https://www.eia.gov/tools/faqs/faq.php?id=27&t=4

In general, in the United States, RINs comprise the following activities[216]:

- Renewable fuel producers generate RINs.
- Market participants trade RINs.
- Obligated parties obtain and then ultimately retire RINs for compliance.

RINs can be traded in two forms:

- Assigned RINs: directly associated with a batch of fuel that travels with that batch of fuel from *party* to *party*. Purchasers obtain *both* the renewable fuel and RINs together.
- Separated RINs: formerly assigned with a batch of fuel, but are no longer assigned to a batch. Purchase only the RIN.

Examples of typical RIN transactions include:

- Generate: when a fuel is produced, a RIN is generated.
- Buy: when an assigned/separated RIN is bought/traded by a buyer from a seller.
- Sell: when an assigned/separated RIN is sold/traded by a seller to a buyer.
- Separate: when a RIN is separated from the fuel to which it was originally assigned.
- Retire: when a RIN is used to demonstrate compliance, or required to be retired for other purposes.

Under this RFS mandate in the United States, refiners are *required* to blend specific volumetric target of renewable fuels each year into their finished petroleum products every year. This target is adjusted annually by the EPA. These volumetric targets are based on the original estimated gasoline demand at the passage of the Energy Policy Acts of 2005 & 2007. At that time, gasoline and diesel demand were increasing every year, causing the target for blending to increase every year, requiring refiners to blend *more* and *more* renewable fuels into gasoline and selling more biodiesel.

But now with a more *subdued* fuels demand environment in the United States, renewable fuels might be entering the "blend wall", or the point at which refiners are required to blend more ethanol into the transportation

---

[216] https://www.epa.gov/renewable-fuel-standard-program/renewable-identification-numbers-rins-under-renewable-fuel-standard

fuel supply than can be supported by the demand for E10 gasoline (gasoline containing 10 percent ethanol by volume[217]). RINs prices and availability have remained volatile due to regulatory uncertainty on the EPA[218].

Fuel Taxes

Fuel taxes in the United States currently average about $0.47 per gallon across the several states[219]. According to the Organization for Economic Cooperation & Development or the OECD, the average fuel tax per gallon around the world is $2.62 per gallon[220]:

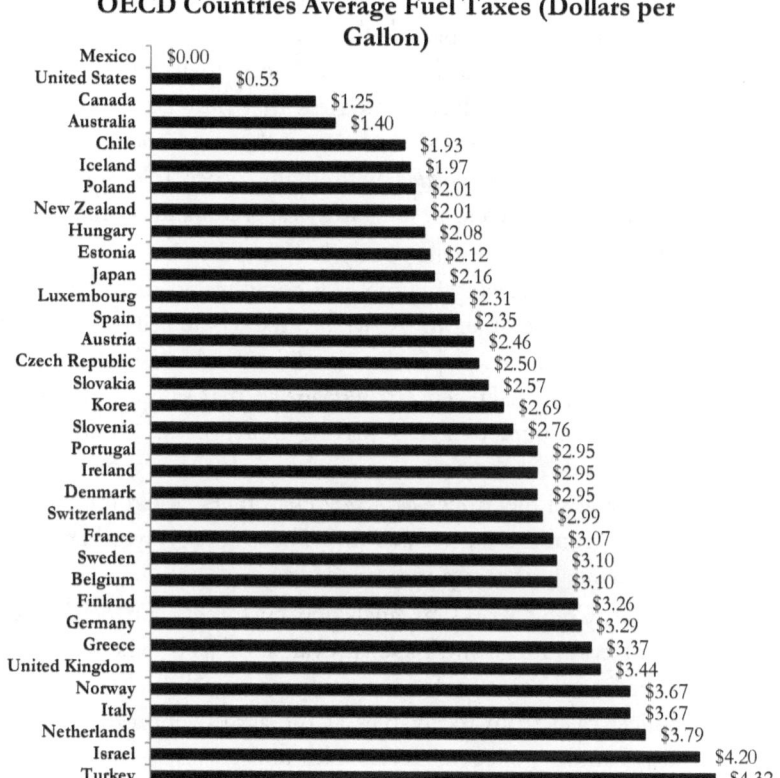

OECD Countries Average Fuel Taxes (Dollars per Gallon)

[217] https://rbnenergy.com/will-rin-and-stimpy-dodge-the-ethanol-blend-wall-in-2013
[218] Western Refining 2016 Form 10-K, page 22
[219] https://www.eia.gov/tools/faqs/faq.php?id=10&t=10
[220] https://taxfoundation.org/how-high-are-other-nations-gas-taxes/

Roles in a Marketing & Retail Organization

There are several roles involved in a Marketing & Retail Organization:

- Marketer or sales representative: The petroleum marketer is a key role in a marketing organization. These marketers are responsible for selling fuel, typically from a wholesale perspective, to retailers or other companies across several locations. Marketers are usually organized into terms of sales areas or territories, for which these marketers are responsible for developing deals or contracts with retail marketing companies to buy fuel from the company through a variety of contracts that can be from monthly contracts to multi-year contracts that span one site or multiple sites. These marketers need to develop deep expertise about their sales territories and need to convince these customers that purchasing fuel from their company is more advantageous than competitors. Marketers usually work *hand-in-hand* with many other roles in a marketing organization, such as pricing analysis, credit and promotions.

- Pricing Analysis: Pricing analysis is another key role in marketing organization. Pricing for fuel is a complicated *science* or more often than not more of an *art*. At what price would selling the company's fuel make customers purchase *additional* volumes? Or at what price would it not make sense to engage in that sale? Under what basis should a contract have a pricing benchmark? Should it be sold below that benchmark price or above? If the organization lowered or increased fuel prices *how many* customers would the company *lose* or *gain* from these price actions? These and many other questions a pricing analysis group tries to answer.

- Credit: Credit works by analyzing customers' financial statements and credit ratings and issue an evaluation as to whether a customer is *creditworthy* enough for the company to sell fuel to and expect to collect cash from receivables at the agreed upon date. This group also handles collecting receivables and ensuring customers pay the company on time since payment terms for fuel sales are very short cycle.

- Promotions or Advertising: this group also works very closely with marketers in trying to understand end-customers need and pricing sensitivities. This group might work in running advertising campaigns to promote a particular company's brands of fuels and partner up with other companies in offering promotions such as price discounts if a certain credit card is used to purchase fuel. This group is also responsible for administering loyalty cards that let customers receive some sort of benefit for being repeat customers for a particular brand.

- Customer relations, customer setups and transactional accounting: This group is in charge of setting up customers in the company's systems as well as ensuring wholesale customers are satisfied and address any complaints that relate to billing issues, fuel quality, rebates or other areas.

- Scheduling/Dispatching: Scheduling or dispatching groups are there to ensure customers can pick up product at the designated transfer point at the *right time* and have the available inventories to meet product demand. They also ensure any logistical disruptions are handled so that the company's contracts and customers are not impacted from these disruptions.

- Retail employees: if a company has company-owned & operated sites, these employees work at the service stations and provide a variety of services at stations, such as operating point of sales, maintaining these retail sites. These are typically the most end-customer facing employees a company may have.

Supply & Trading

Typically, most oil & gas companies, whether they operate in upstream, midstream or downstream, have a substantial presence in the supply & trading (S&T for short) of hydrocarbon commodities. These organizations within a downstream company are usually organized along commodity and sub-commodity lines as well as between supply & trading groups. An S&T organization provides a critical link between the market and a company's hydrocarbon operations, and in particular between the Refining & Marketing operations of a downstream company.

Supply

The overall goal of the supply department of a Supply & Trading organization can be summed up by the following objectives:

- Secure the *best prices* for a company's production of hydrocarbons, whether they are crude oil or natural in the case of an upstream or integrated company or for refined products or NGLs in a downstream company.

- Purchase required crude oil, feedstocks and other hydrocarbon products at the *lowest possible price*, so that a refinery or marketing organization can improve margins.
- Guarantee production outflow from the company's refineries and guarantee consistent refined product supplies to the marketing organization.

Trading

One of the key concepts in all types of trading is the concept of arbitrage. Arbitrage is simply the practice of taking advantage of a price difference between two or more markets in order to profit from these discrepancies[221]. An example of an arbitrage opportunity in the crude oil markets might be that the *same* crude oil is selling in location A for $45 per barrel while in location B is selling for $52 per barrel. An arbitrage profit could be made by buying in location A for $45, transporting that crude oil for $2 to location B and then selling it there for $52 and net a profit of $52-$45-$2 which equals $5 per bbl. As more market participants discover this arbitrage and continue to execute on this arbitrage opportunity the price in location A will tend to *rise* and will tend to *equal* the price in location B *minus* the transportation to get that barrel there.

Arbitrage opportunities tend to arise since information is *asymmetrical* in the sense that it is impossible for all market participants to have the *same* level of information about a particular commodity, location or particular local needs. Arbitrage opportunities tend to happen usually in markets with low *liquidity* or low level of buying and selling volumes. As more market participants take place in buying and selling a commodity, prices will tend to be more aligned. Liquidity can be defined as how *quickly* and *inexpensively* an asset or commodity can be converted into cash. Assets or commodities, like real estate, that can only be sold after a long and possibly exhaustive search for a buyer is considered as *illiquid*[222].

The key role of trading is *how to discover* these arbitrage opportunities and *execute* buy or sell deals in order to *capture* these opportunities. Trading in a market, or the buying and selling of a commodity, increases the *liquidity* so that as a market develops and becomes more traded, *less* arbitrage opportunities will tend to exist over time.

---

[221] https://www.merriam-webster.com/dictionary/arbitrage
[222] https://www.thoughtco.com/definition-of-liquidity-1146123

Price Differentials

For both Marketing & Trading, the *price differentials* between what they buy and sell, is critical, not necessarily the *outright absolute* price of a commodity. In the downstream business, differentials are *much more* important than the absolute value of what the commodities sell for. A wholesale marketing company could generate higher earnings at *lower fuel prices* as long as the price *differential* between say the distribution terminal and the retail price of fuel stays high, the absolute level of price does not impact as much.

Price differentials are key, which is why the following are the three laws of energy markets[223]:

- First Law: Energy wants to move from a *lower-value* market to a *higher-value* market
- Second Law: If there is no capacity or other logistical constraints, the price differential between those *lower-value markets* to *higher-value markets* will narrow until it equals *transportation costs* between those markets.
- Third Law: lack of transportation capacity invalidates the Second Law.

More importantly, price differentials are *easier* to predict than the *absolute* value of a commodity. Forecasting the *absolute* level of a commodity such as crude oil is a very complex and error prone endeavor that depends on many factors, such as geopolitics, the value of the dollar, the state of the world economy and many factors. Meanwhile, the factors influencing price *differentials* can be predicted with far fewer variables and simple analysis. Factors that influence differentials are primarily a function of local energy markets that is supply, demand flows and infrastructure, all fundamental factors that can be measured and modeled[224].

When price differentials are high, the trading business does well since it is able to capture more arbitrage opportunities in the market and profit from these. The trading business will try to capture this arbitrage in the market until more and more players reduce the differential to reflect transportation costs long term. For example, if there are significant arbitrage opportunities in purchasing discounted crude oil in Midland, TX and ship this crude oil where it is more highly valued in the Texas Gulf Coast, traders might find alternative transportation and over time encourage midstream companies to

---

[223] The Domino Effect, by Rusty Braziel, Kindle Edition, locations 3711-3787
[224] Ibid

build new pipelines that once built will *reduce* these differentials to reflect the cost of transportation. An example of this is the new Enterprise Midland to Sealy pipeline, which was announced in April 2015, will become operational in late 2017 or early 2018 and is expected to reduce these differentials:

> *Enterprise Products Partners LP (NYSE: EPD) said April 30 it plans to develop a new pipeline to transport crude oil and condensate from Midland, Texas, to the Houston area. Enterprise has long-term agreements to support the 416-mile, 24-inch diameter pipeline, which would connect Enterprise's Midland terminal to its storage facility in Sealy, west of Houston. It would then link to Enterprise's ECHO terminal through an interconnect with the 36-inch Rancho II pipeline, which is expected to be in service in July[225].*

There are several types of differentials associated with commodities:

- Location differentials, as the name implies, are price differentials associated with the *receipt* and *delivery* of commodities in different *locations*. In the crude oil business, the exact same crude oil is a lot more valuable when is available closer to a refinery than where it is out in the field. Therefore, the price differentials would, over time, reflect the cost of transporting that crude oil from that *less valuable* location closer to the refinery.

- Quality, not all crude oils or products are the same; therefore, there are quality differentials embedded in the price. As an example, a *light sweet* crude oil would tend to command a higher price than a *heavy sour* crude oil, since the light sweet crude oil is higher quality.

- Time, in a normal market, commodities or all goods for that matter, are valued *more* today than in the future. In the trading world, this type of normal market is called *contango*, whereby the *prompt* or current month valuation on the futures market is valued *higher* than for a future month delivery. This makes sense since there are risks involved in a future delivery vs. delivery of a good now. Many times in the commodities market, a future month delivery commodity is *valued* more than the current market and that is called in trading lingo *backwardation*. When markets are in *contango*, the majority view expects prices for that commodity to *rise*, while markets in *backwardation* are expected to have *lower* prices in the future.

---

[225] http://www.bizjournals.com/houston/morning_call/2015/05/enterprise-plans-new-texas-pipeline-amid-oil-slump.html

Crude Oil Supply & Trading

The main goal of a crude supply organization in a refining company is to acquire, at the *lowest cost* possible, the best crude oil that will provide the *highest yield* of desired petroleum products to a refinery. To that extent, a crude oil supply organization has to be heavily engaged with a company's refineries in order to better understand the following:

- Existing configuration of company's refineries and which crude oils or type of oils a refinery can run versus the other. There are different oil quality requirements that a refinery is looking for, initially the API gravity scale and sulfur content. As previously covered in the refining chapter, refineries have different equipment that allows them to run certain types of crude oil, with the most complex refineries being able to run more types of crude oils.

- Model or simulate the projected refined products production or *yields* from each of the different crude oils and assign prices to each product.

- Assist a refinery in selecting the best available crude oil based on that refinery's capabilities and calculate a "netback" crude price, which is then used to price a certain crude oil at a discount or a premium against a benchmark crude oil like WTI or Brent.

Calculating Crude Oil Value to a Refinery

This is an activity that involves groups like optimization, refining operations, market analysis and trading in order to maximize margins at the refinery and purchase the most suitable crude oil. The following table provides an example of how to calculate what the value and products that could be produced from a particular crude oil:

Product	Value of Product per Barrel $/BBL	Yield	Value of Product per *Crude* Barrel
Propane	$37.85	0.037	$1.40
Gasoline	$62.58	0.475	$29.73
Jet	$65.47	0.125	$8.18
ULSD	$68.52	0.284	$19.46
Residual Fuel Oil	$38.96	0.025	$0.97
Petroleum Coke	$4.50	0.057	$0.26
Refinery Gate Value		1.003	$60.00
Refining Cost			($4.78)
Transportation & Logistics			($2.50)
Crude Netback Price			$52.72

The information above is based on *crude assays* for each product as well as the refinery's particular configuration, which provides the expected yield for each major product for this particular crude oil. Each barrel of crude oil has a *particular* yield of products based on specific units to a refinery. Prices per barrel for each product are then multiplied for the expected yield to arrive at the value of product per *crude* barrel. These values are then added up to arrive at what it's called the *Refinery Gate Value*. The Refinery Gate Value is the expected value a refinery would expect to receive from selling each product in terms of what it generated per *each* barrel of *crude oil*.

This refinery gate value is then reduced by what it costs to the refinery to produce all those products in terms of each crude barrel. This value is then further reduced by subtracting out transportation & logistics costs, which are the costs the refinery would incur by transporting that barrel of crude oil from the market center or delivery point it purchased from.

The end result of this calculation is the final answer that crude purchasers or traders are looking for, the *Crude Netback Price*. The Crude Netback Price is the price a refinery would be *willing* to pay to achieve that desired level of profitability or margin. In other words, this is the *maximum* price the refinery would be willing to pay for this particular type of crude oil and any price paid below this netback value would be considered an upside to a trading organization.

The same iterative calculation would have to be performed each time a new crude oil is assessed as well each time each product price or other condition changes. Due to the complexity of these calculations and how often they need to be run, many companies have designed complex *Linear Programming* models that can be run from either complex spreadsheets or by using more robust computer power with specialized software applications[226].

How Crude Oil is Priced

Currently, West Texas Intermediate (WTI) and Brent are both the global benchmark contracts for crude oil pricing. The WTI futures contract trades on the New York Mercantile Exchange (now owned and operated by CME group or Chicago Mercantile Exchange) and the Chicago Board of Trade and is the most actively traded energy product in the world. A couple of key facts about this contract[227]:

- Liquidity: WTI is the most liquid global energy benchmark, trading nearly 850,000 futures and options contracts *every day*.

---

[226] http://www.argusmedia.com.br/methodology-and-reference/~/media/E73A4C20016F4F9D92B573180BB1428A.ashx
[227] http://www.cmegroup.com/trading/energy/light-sweet-crude-oil.html

- Open interest[228]: Open interest has exceeded 3 million lots, which equates to more than 3 million barrels per day.

- Significant physical market reference, serving as the *benchmark* for approximately 10 million barrels per day of North American production.

Other crude oils *tend to* follow the pricing trend of WTI and Brent crude oils and would trade at *discounts* or *premiums* depending on the oil's quality. For example, a crude oil with an API of 30 would sell for a discount in comparison to WTI, which has an API of 39.6. The same would apply for physical delivery or location discounts, in other words, a crude oil to be received close to a refinery is *more* valuable than one being delivered to a pipeline or terminal far away from any valuable market center.

Crude oils similar in quality such as Brent and WTI tend to be *generally* sold for a similar price[229].

As mentioned before, lighter and sweeter crude oils would tend to have a higher market price than heavier and sourer crude oils. The table below shows a sample crude oil bulletin from Plains Marketing LP for the month of March 2017[230]:

Crude Name	Posted Price	Gravity Range	Adj. Scale
West Texas Intermediate (WTI)	$45.6452	40.0-44.9	Less $0.02 per barrel for each full degree API below 40.9°; less $0.015 per barrel for each 0.1 degree API above 44.9°
Oklahoma Sour	$33.6452	40.0-44.9	Less $0.015 per barrel for each 0.1 degree API below 40.0°; less $0.015 per barrel for each 0.1 degree API above 44.9°
South Texas Heavy	$39.6452	29.0-44.9	Less $0.02 per barrel for each full degree API below 29.9° to 19.9°; less $0.015 per barrel for each 0.1 degree API above 44.9°

---

[228] Open interest is the total number of options and/or futures contract that are not closed on delivered on a particular day. From http://www.investopedia.com/terms/o/openinterest.asp

[229] Since 2011, WTI and Brent have traded at significantly high differentials.
http://www.bloomberg.com/news/2014-05-06/wti-oil-rises-on-forecast-cushing-weeks-from-emptying.html

[230] https://www.plainsallamerican.com/getattachment/609ea007-82f5-47ad-a995-0c20695f8ede/March-2017-Recap.pdf?ext=.pdf

Refined Products Supply & Trading

The market for refined petroleum products is similar to the market for crude oil in the sense that similar activity takes place:

- Refined products trades are transacted in both the *physical* and *futures* markets.
- These products can be bought and sold in organized exchanges, like the New York Mercantile Exchange, the Chicago Mercantile Exchange and many others.
- Products usually trade in terms of *premiums* or *discounts* to the market major and product indexes.

Petroleum products, like crude oil, are global commodities, and as such, their prices are determined by supply and demand. These prices reflect the many interactions of many buyers and sellers, each with their own expectations of supply & demand, with these interactions occurring both in the physical and future markets[231].

Exchange & Purchase Agreements

One of the reason for this is that the fact that a lot of supply & trading activities occurs between different oil companies through are what called *purchase agreements*. For example, Valero may have requirements to supply reformulated gasoline in the Southwestern US area, but they do not have a refinery, they could contact a supply representative from a local refiner in order to supply Valero's need for fuel in that specific area of the company.

The same logic applies with the concept of *exchange agreements*. Exchange agreements provide for the delivery of refined products by the downstream company to third party companies at *specific locations* in exchange for delivery of a similar amount of refined products to the downstream company to other oil company at another location. These exchange agreements in effect *decrease* transportation costs by not having to ship volumes from one point to the other as well not have to re-formulate gasoline or other products. Say that Shell requires volumes to sell to its wholesale customers in Los Angeles, but their supply is located in Houston. On the other hand, Chevron has excess supply of product in Los Angeles and does not have available supplies in the Houston. Shell and Chevron could enter into *exchange agreement* whereby Chevron would supply Shell with gasoline in the Los Angeles market and in return Shell would do the same for Chevron in

---

[231] API, "Understanding Crude Oil and Product Markets", page 6

Houston and that they would settle any product or location differentials in *cash*. This is an effective technique in the supply & trading business that facilitates meeting each company's requirements for products at different locations. The same concept would apply as well with *time exchanges*, which are exchanges that are not necessarily *location* driven, but more to balance out excess supply in one period versus excess demand in another period for a particular company.

These agreements help minimize transportation costs, optimize refinery utilization, balance refined product availability, broaden geographic distribution, and provide access to markets not connected to a company's refined-product pipeline systems[232].

Regulations

Refined products are very dependent on local supply and demand dynamics. In many areas, due to environmental regulations, each location may require a specific type of gasoline. Such is the case with CARBOB or the California version of regular gasoline. In additional to federal EPA mandates for gasolines, many jurisdictions in the U.S. mandate different formulations for gasoline, limiting product flows between two locations. Another regulation reducing product flow between the different regions is the high cost of U.S. *built* and *operated* ships, which is mandated by the *Jones Act of* 1920, which requires all product shipments between two U.S. ports to be made with *Jones Act* vessels. Jones Act vessels tend to be significantly higher cost than international ships[233], typically up to 5 times as high as a foreign ship. This limits the ability to effectively ship refined products between two U.S. locations and thus might incentivize exporting that product abroad.

Refined products are bought and sold all across the world, with several key futures contracts being used as reference point including:

Name of Contract	Exchange	Description & Specifications
NY Harbor ULSD Futures	NYMEX	Ultra-low Sulfur Diesel, 15PPM sulfur maximum, with a minimum API gravity of 30. Based on the product specifications for the Colonial Pipeline. Delivery point is New York Harbor
RBOB Gasoline	NYMEX	RBOB or "Reformulated Gasoline Blendstock Oxygen Blending (RBOB). Delivery point in New York Harbor. 87 octane rating RBOB gasoline after blending with 10% denatured ethanol.

---

[232] Valero 2016 Form 10-K, page 8
[233] https://comptroller.texas.gov/economy/fiscal-notes/2016/january/jones.php

Gasoline for example would be traded in terms of a *premium* or *discount* relative to the RBOB gasoline futures contract. For example, the same 87 octane gasoline could be bought at a discount of 7 cents per gallon versus the futures contracts. The concept of differentials simplifying buying and selling since it is looking at the *spread* between one delivery point and the other delivery point, not at the absolute price level of gasoline.

Exports

For the first time since 1949, the United States has now become a *net exporter* of refined petroleum products, creating significant opportunities in markets abroad. The chart below provides an illustration of these higher export volumes[234]:

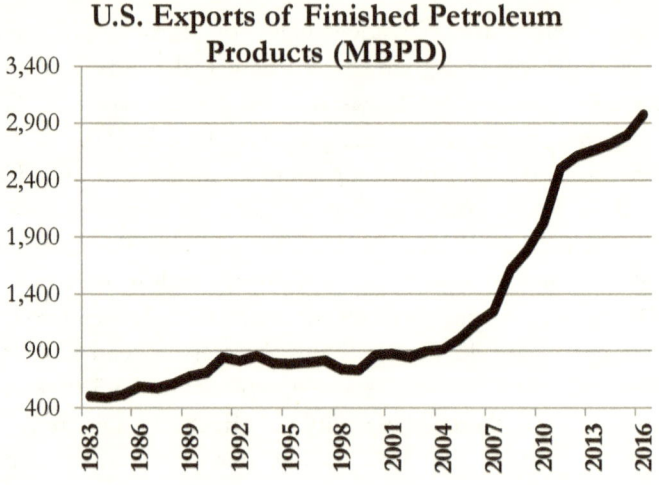

U.S. Exports of Finished Petroleum Products (MBPD)

Natural Gas Liquids Supply & Trading

Natural Gas Liquids or NGL's for short, are valuable hydrocarbon components that can be extracted from *wet* natural gas or be produced from refining processes, in particular cracking and distillation. NGLs around the world are traded in terms of gallons or barrels. Since there are five natural gas liquids (ethane, propane, iso-butane, normal butane and pentanes plus), there is no single price for a combined "NGL barrel" per se. The five NGLs have distinct markets and volatilities associated with them.

---

[234] Source: EIA

In the United States, there are two major market centers for NGL supply & trading:

- Mont Belvieu and Louisiana have the world's largest fractionation centers, with these two locations providing interconnection with many pipelines and having petrochemical customers connected with the distribution systems on these locations. Moreover, since the U.S. Gulf Coast is the largest refining center in the world, having close proximity and the infrastructure to deliver purity NGL products to be processed into a variety of feedstocks and blending opportunities increases the number of customers interested in buying NGLs. Mont Belvieu and Louisiana areas are estimated to have around 3.12MMBPD of fractionation capacity[235], providing access to the most liquid markets for NGL supply or trading in the world.

- Conway, KS, which is home to several fractionators, has large storage capacities with more than 30 active underground caverns, rail loading facilities for purity products as well as connection to petrochemical facilities[236].

- There are other NGL trading centers in North America, including Sarnia, Ontario, Detroit, MI, Hobbs, NM and other locations.

Outside the United States, NGL pricing is *less structured* and it is typically more tied to oil pricing, although this has started to change as the United States has become a major LPG exporter over the last couple of years[237]. It is expected as LPG exports from the United States increase that LPG market prices will tend to converge and long-term become more in sync to U.S. NGL pricing.

Before the shale revolution in the United States, NGL prices used to track the price of crude oil very closely. After massive increases in NGL production in the U.S., NGL prices became disconnected from crude oil prices and started to trade more independently of the fluctuations in crude prices.

Role of NGL Supply in a Downstream Company
The refining process is both a *producer* and *consumer* of NGL, therefore the role of an NGL supply group is to both supply certain NGLs needed for

---

[235] https://www.oilandgas360.com/outlook-positive-u-s-ngl-production-exports/
[236] http://co.williams.com/operations-2/west-operations/mid-continent-fractionation-storage/
[237] Enterprise Products Partners 2017 Analyst Meeting, slide 3

refinery processes, such as isobutane for gasoline blending, and sell other NGLs that are not needed at a refinery, like for example propane.

Natural Gas Supply & Trading

Natural gas is bought and sold in terms of MMBTU or million British Thermal Units. What is a BTU? A BTU is the amount of energy required to heat one *pound* of water by one *degree* Fahrenheit. To put it in perspective, one single BTU generates about the same energy released as the burning of a single wooden match.

Natural gas is bought and sold in terms of heating content or energy so that all the various hydrocarbon components are reflected in the price (primarily methane and ethane, but can include other heavier NGLs). If natural gas were to be bought and sold in terms of volume, the higher value components which have a higher BTU content would not be reflected in the price. The second, and perhaps the most important reason, is that there are BTU specifications across the entire gas value chain, starting at the "burner tip" of a typical gas range going all the way to a gas gathering pipeline. The reason these BTU specifications are important is that all appliances or equipment that use natural gas in a certain country or area are made for a very specific composition of natural gas. If higher BTU content natural gas were allowed into a transmission pipeline, this could cause an array of problems, from liquid buildups in pipelines, damages to equipment burners, or more importantly, creating a safety hazard for the many stakeholders involved.

In terms of markets and pricing, natural gas is *largely* a regional market, and unlike crude oil, there is no single global market for natural gas. Different areas around the globe will have substantially different prices from one another[238]. While in the United States, natural gas might be selling for $3 per MMBTU, in Europe natural gas might be selling for $6 per MMBTU while in Japan it might be selling $8 per MMBTU. The United States has enjoyed some of the lowest natural gas prices in the developed world for several years now, thanks largely due to increased hydrocarbon production in the U.S.[239]. Due to the fact that a pipeline or an LNG plant has to be built *first* in order to transport natural gas from one place to the other, natural gas is expected to remain a regional market for the next several years[240].

---

[238] http://www.eia.gov/todayinenergy/detail.cfm?id=3310
[239] http://www.eia.gov/dnav/ng/hist/n9050us2a.htm
[240] http://www.bloomberg.com/news/2014-09-30/u-s-gas-boom-turns-global-as-lng-exports-set-to-shake-up-market.html

Natural gas, once processed by a gas plant, is a homogenous product adjusted for BTU quality or heating content. In contrast with crude oil, crude oil is not a homogenous product, but as mentioned earlier, has to be adjusted for quality factors such as gravity and sulfur content.

As discussed in chapter three, natural gas is a key component in refining operations, both as a *feedstock* and an *energy* source. Typically in a downstream company, there would be a dedicated group purchasing and possibly trading natural gas. The main focus of this group is usually to find reliable natural gas supplies for the company's refineries and be able to procure natural gas and electricity at the *lowest* possible cost so that it improves the competitiveness of the refinery.

Roles in a Supply & Trading organization

There are different roles or position within a supply & trading organization in a downstream company:

- Origination & Business Development: this group of people is responsible for *connecting* customers with the supply & trading organization and can work with variety of commodities or developing new businesses and optimizing assets like fractionation, transportation of hydrocarbons, export terminals and many others. For example, they might connect a power plant that requires natural gas supply to run their power plant and then work with traders, schedulers and other positions in executing these deals. Originators work in particular in connecting markets that are typically not organized or structured via standard platforms and exchanges[241].

- Traders: actively buy and sell hydrocarbon commodities, primarily on exchanges like Intercontinental Exchange, Chicago Mercantile Exchange and many others. They work daily with traders from other companies and connect with them to sell for example excess crude oil a refinery might not need or trading additional gasoline volumes that a wholesale marketing customer might not need.

- Optimization: Optimization works very closely with refineries in particular, tweaking the so-called Linear Programming models refineries run in order to select the best and most economical crude oil a refinery can run. For example refinery A might like to run a cheaper crude oil because certain of the by-products that refined

---

[241] http://www.engineeringbecause.com/news/engineering-europe/69/interview-with-an-originator-beyond-the-trading-desk-at-bp

with this crude oil have increased in price, therefore the refinery's crude input could be optimized in future weeks by switching to a *lower cost* crude oil.

- Market Analysis: Market Analysis works in providing price inputs, market analysis and insights to traders and business development. This group careful analyzes the commodity markets to find gaps or opportunities for arbitrage that could then be executed by traders. They also work very closely with other groups within a downstream company in providing market commentary and forecasts, which are sought after, particularly in turbulent markets.

- Schedulers: Schedulers, as the name implies, are responsible for *scheduling* the movement of crude oil, natural gas, NGLs and products across different modes of transportation, such as pipelines, ships, barges, trucks and others. Schedulers can participate in tasks such as making sure a refinery has enough crude oil to operate and enough inventory capacity available so that the facility can continue to operate as required. They also assist with customers in nominating and confirming their activity for the following month and ensuring all products flow without interruption through the different modes of transportation required. Depending on the commodity being scheduled and the mode of transport, schedulers' roles can vary in terms of responsibilities. Schedulers are usually expected to be available 24/7 to assist with any unexpected interruption.

- Credit: credit plays a vital role in a Supply & Trading organization ensuring that the customers which the organization transacts are creditworthy *before* entering into any deals with them as well as ensuring customers are making prompt payments. The role and impact of credit has increased, especially after the financial crisis in 2008 in ensuring consisting cash collections from customers. Since all deals are usually paid after delivery and payment terms may vary by commodity (i.e. crude oil in the U.S. is typically on 30-day payment terms, while gasoline is usually less than 5 day payment terms), it is important to be conducting business with customers that have the ability to pay and will not impacting a company's cash flow by delaying payment or simply not paying for a delivery of a commodity.

- Risk: the role of risk is to monitor the organization's exposure with counterparties and prevent unexpected losses in any trading activity that the organization engages in. Risk is also very critical since it calculates traders' *Profit & Loss* or P&L for short. Risk evaluates activities and deals taking place and monitors the risk that the company is being exposed to with different tools, such as *Value-at-Risk*, which estimates how much a set of deals might lose their value in any given time period. Risk is usually associated with compliance activities, such as qualifying customers or suppliers that are prevented from trading or that are based in certain countries that are sanctioned by governments or international agencies.

- Crude oil purchasers: crude oil purchases play an ever increasing role in a downstream company's acquisition of crude oil at the lease or what is commonly called as "lease-purchased" crude oils in the United States. One of the key aspects of purchasing crude oils at the lease is that it offers companies a discount over buying crude at market center points and it also allows the company to buy *unblended* or *uncommingled* crude oils which have a distinct and consistent quality. These unblended crude oils can be run more predictably and reliably through a refinery produce consistent yields of products, in comparison to blended crude oils or what some refineries call *dumbbell* crude oils[242].

- Back office processes: These include primarily accounting, information technology and customer/vendor setup processes. These roles ensure that a supply & trading organization has the adequate software & IT support, that transactions are correctly processed, cash is collected and applied to customers, vendors are paid correctly and all the information flow from deal to financial statements are properly accounted for.

Marketing, Retail & Trading Key Metrics Overview

Metrics such as ROCE, dividend yield, ROE, capital expenditures can also be used to analyze a downstream company or segment. In this chapter, the following metrics are introduced:

- Total refined product sales
- Total number of stations
- Fuel volume sales
- Fuel volume per station
- Gross margin per gallon or liter

---

242 Refineries across the United States

- Merchandise sales as percent of total
- Earnings per barrel
- Cash per barrel

Total Refined Product Sales

Total refined product sales is a volume-based metric, similar to total throughput volumes in refining, but that is used for a downstream company's marketing operations. A company could have more refined product sales than their own refinery throughput volumes since they could purchase products from other third party refining companies. Exchanging or purchasing products from other refining companies makes sense since a company could very well have marketing operations in an area of the world and not have *company-owned* refineries to supply their marketing operations. The same situation could be if the company has a refinery in one area of the country but doesn't have marketing operations, it will therefore sell those refinery volumes to a third party company.

> *Refining Company ABC's refineries had total throughput volumes of 100MBPD in 2016 while ABC's total refined product sales were 250MBPD. Therefore it can be inferred that ABC purchased 150MBPD of products from a third party refinery to fulfill marketing operations commitments. The other inference that can be made is that ABC's marketing operations are quite substantial.*

As a side comment, please note that the practice of refined product exchanges between companies is very commonplace. For example motor gasoline bought from an ExxonMobil branded service station may very well have come from a Shell refinery and vice versa. Usually, the main differentiator from one brand of gasoline to the other would be the different additives added to the gasoline at product, storage terminals or even at the service station.

Total Number of Stations

Total number of stations provides an insight into the retail presence of a marketing organization. As discussed previously in the chapter, these stations can be a part of several types of arrangements, such as being *company-owned & operated, lessee/lessor relationship* or *dealer owned & operated.*

> *Petroleum Marketing DEF had 200 company-owned & operated stations in the United States, while the company had 300 dealer owned & operated stations in Europe. Therefore, the company's retail stations are a total of 500.*

Fuel Volume Sales

Fuel volume sales, similar to refined product sales, provides an insight into how big the retail marketing operations of a downstream company are. Fuel volume sales are *different* from refined product sales in the fact that a company may purchase third party volumes outside of its refining or marketing operation if it makes more sense. For example, if a refining company has supply in Boston, but also retail operations in the West Coast, but no refining supply, it makes sense for the company to purchase wholesale volumes from third parties in order to able to supply its retail or store commitments in that area.

> *Integrated Downstream Company IAC had refined product sales from its operations of 250MBPD, while its fuel volume sales at its marketing/retail operations were 450MBPD. Therefore we can conclude that this company purchases volumes outside of its refining system to supply the company's retail commitments.*

Fuel Volume per Station

This metric provides an idea of how *efficient* an average service station is in comparison to peers. This metric also provides an insight into the different fuel retail dynamics, such as certain regions of the world having stations with a *relatively* low number of fuel sales per site, like those in Latin America vs. gas stations in the United States which tend to sell more fuel per site.

> *Marketing Company DEF had 200 stations in the US, 100 in Canada and 200 stations in Germany, indicating total stations of 500. Total fuel volume sales in 2016 were 110MBPD, indicating average volume per station of 220BPD. The stations in the US sold a combined total of 60MBPD, Germany 30MBPD and Canada 10MBPD, which translate to an average per station of 300BPD in the US, 150BPD in Germany and 100BPD in Canada.*

Gross Margin per Gallon or Liter

The gross margin per gallon or liter measures the revenues *minus* cost of goods sold or margin for every gallon or liter of fuel that is sold. Gross margins can change substantially and typically gross margins are higher whenever you have falling crude oil and refined petroleum products as there is a *lag* effect between when prices start to fall and when stations deplete their inventory.

> *Marketing Company DEF had total fuel revenues of $10MM on 5MM gallons that it sold. The company's total fuel purchases were $8MM, indicating that gross margin was $2MM. Therefore, the gross margin per gallon was $0.40.*

Usually, gross margins are in the $0.10 to $0.20 per gallon in the United States, but are typically higher in Europe and other regions of the world.

Merchandise Sales as Percent of Total Sales

Merchandise sales have a higher margin than regular fuel sales, with merchandise averaging gross margins of 30-40%[243] in comparison to fuel margins, which in the United States have averaged around 6% over the past five years[244]. This metric is calculated by taking the total merchandise sales and dividing it by total sales.

> *Marketing Company DEF sold $3MM in merchandise sales in 2016 while fuel sales were $30MM. Therefore, the company's merchandise sales as a percent of total is now $3MM divided by $30MM equals to 10%.*

Earnings per Barrel

Similar to the Refining chapter, earnings per barrel is a financial measure of how *profitable* a marketing company is on a GAAP *accrual accounting basis*. A company with a high quality portfolio of marketing contracts (high margin), effective trading operations, good portfolio of retail stores) that runs efficiently, controls costs and increases revenues would tend to have higher earnings per barrel than its competitors. Similar to chapter three, these earnings may have to be adjusted for special items.

This metric is calculated the following way:

- GAAP earnings (adjusted for non-cash or special items) for the period *divided* by total fuel volumes sold.

> *Marketing Company DEF had earnings of $200MM, which included an asset impairment of $40MM and had total fuel volumes sold of 10 million barrels for the period. Therefore, adjusted earnings were $240MM, which is then divided by total fuel volumes sold for the period of 10 million barrels, which equals to adjusted earnings of $24 per barrel.*

---

[243] CST Brands: Merchandise and Service
[244] 2016 NACS Retail Fuels report page 11

Cash per Barrel

Cash per barrel is a metric that measures of *how much* cash is being generated by a marketing company on a per *barrel* basis. Total cash (the numerator) is calculated as follows:

- Marketing GAAP Income *plus*
- Adjusting or special items *plus*
- Depreciation & Amortization (D&A)

Similar to earnings per barrel, the denominator for this metric is total fuel volumes, usually stated on a yearly or quarterly basis.

> *Petroleum Marketing Company CEF, in 2016, had $300MM of earnings, D&A of $150MM, while total fuel volumes were 15MM barrels. Cash per barrel for the period is thus $30.*

Business Cycle in Marketing & Trading

As discussed in the price differentials section and similar to refining, both marketing and trading depend not so much on the absolute price level of the commodities being purchased and sold, but more on the price differentials. Below is a chart of the EIA representing the historical price differential between WTI and Brent[245]:

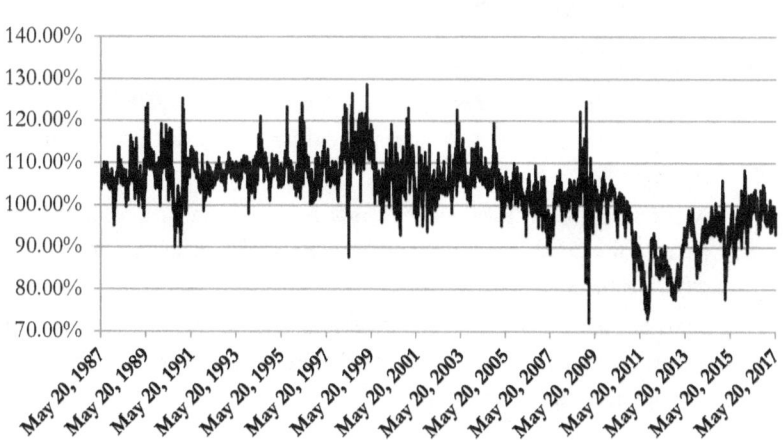

WTI as % of Brent since 1987

---

[245] https://www.eia.gov/dnav/pet/pet_pri_spt_s1_d.htm

For many years before the late 2000's, WTI used to be more expensive than Brent crude oil despite being very similar oils in quality. The main reason for this was the *declining* US crude oil production, which placed inland price benchmarks at a *higher price* than overseas crude oil. This was because Brent crude oil could be more easily transported in large vessels while transporting an imported crude oil all the way to the delivery point of Cushing, OK from the Gulf Coast was more expensive. What began to happen in the mid to late 2000's with the shale revolution, U.S. crude oil production started to increase and not have a corresponding market outlet (before 2016, exporting crude oil out of the US was simply not allowed) which caused an *oversupply* of WTI crude oil versus declining Brent production[246] and this shifted WTI from a *premium* crude oil to a *discounted* crude oil in this space from the early 2010's until the mid-2016's. As pipeline infrastructure that transported barrels from Midland, Texas to the Gulf Coast became operational plus the subsequent lifting of the export ban on crude oil, the price differentials between WTI and Brent started to return back to more historical normal levels. Coupled with this newly built infrastructure was the fact that U.S. crude oil production was declining after the decline in oil prices that started in mid-2014. The result was that the WTI to Brent price differential started to narrow and went back to historical normal levels around price parity.

Why Invest in Marketing?

The marketing business provides several key advantages:

- Marketing gross margins are fairly stable and in the U.S. for example, have averaged anywhere from $0.11 to $0.38 per gallon, this is contrast to crack spreads, which can fluctuate significantly more.

- Marketing provides a *natural hedge* for declining prices for refined petroleum products, taking advantage of the price lag between *rack wholesale* prices and *retail* prices.

- Marketing, although a *low return* business, generates substantial amounts of cash over time that tends to be more stable than the cash generation from refining.

- Retail stations also provide additional income streams, including merchandise which typically has higher margins than fuel margins.

---

[246] https://media.mhfi.com/documents/201509-brent-crude-on-the-decline.pdf

Chapter V – North America

"If everyone is moving forward together, then success takes care of itself." –
Henry Ford

ExxonMobil Downstream (XOM)

www.exxonmobil.com

"Well done is better than well said" – Benjamin Franklin

Company Overview

ExxonMobil is the largest publicly traded integrated energy company in the world. ExxonMobil is a fully integrated energy company with operations in upstream, downstream and chemicals sectors of the oil & gas industry. With a $374 billion market capitalization at year-end 2016[247], ExxonMobil is the largest *publicly traded* oil & gas company in the world and largest global refiner in the world[248].

ExxonMobil predecessor companies were originally founded in 1870 by John D. Rockefeller as part of Standard Oil group of companies. Standard Oil Company of New Jersey (Esso), Standard Oil Company New York (Socony), Vacuum Oil Company and Humble Oil Company were the primary precursor entities forming today's ExxonMobil. ExxonMobil was incorporated in the State of New Jersey in 1882. The company's early beginnings trace back to being a domestic refiner and distributor of kerosene to becoming the largest publicly traded energy company in the world. On November 30, 1999, Exxon and Mobil merged to form the ExxonMobil Corporation.[249]. In 2009, Exxon purchased XTO Energy for $31 billion, which was one of largest natural gas producers in the U.S. at the time, increasing significantly the company's natural gas reserves in the U.S[250].

Headquartered in Irving, Texas, ExxonMobil is involved in exploration and production of crude oil and natural gas, manufacture of petroleum products and transportation and sale of crude oil, natural gas and petroleum products. ExxonMobil is a major manufacturer and marketer of commodity petrochemicals, including olefins, aromatics, polyethylene and polypropylene plastics and a wide variety of specialty products. ExxonMobil also has interests in electric power generation facilities[251].

---

[247] Based on 4,148 million shares outstanding at year-end 2016 and a year-end closing price of $90.26. Source ExxonMobil's website and 2016 Form 10-K, page 51
[248] http://corporate.exxonmobil.com/en/company/worldwide-operations/locations/united-states
[249] http://corporate.exxonmobil.com/en/company/about-us/history/overview
[250] http://corporate.exxonmobil.com/en/energy/natural-gas/operations/xto-energy
[251] ExxonMobil 2016 Form 10-K, page 1

ExxonMobil is listed on the New York Stock Exchange (NYSE) and trades under the ticker symbol XOM.

At year-end 2016, ExxonMobil had more than 71,000 employees, primarily located in the United States[252].

Areas of Operation

ExxonMobil's operations are divided into three operating segments, Upstream, Downstream and Chemicals:

- Upstream is engaged in the exploration & production of oil and natural gas, including LNG operations.
- Downstream is engaged in the refining and marketing of petroleum products. ExxonMobil is the largest refiner in the world.
- Chemicals is engaged in the manufacture and marketing of petrochemicals. Exxon Mobil Chemical is one of the largest chemical companies in the world and the largest major chemical manufacturer in the United States[253].

Downstream

ExxonMobil's downstream segment manufactures and sells petroleum products. The refining and supply operations include a worldwide network of manufacturing plants, transportation systems, and distribution centers that provide a range of fuels, lubricants and other products and feedstocks to its customers around the globe.

ExxonMobil is the world's largest refiner, and as of 2016, had an interest or operated 22 refineries in the world with a crude oil processing capacity of 4,907MBPD on a net basis or 5,540MBPD on a gross basis (atmospheric distillation)[254]. ExxonMobil had refining operations in North America, Europe, Asia-Pacific and the Middle East.

North American Refineries

In North America, ExxonMobil had eight refineries with a total crude distillation capacity of 2,146MBPD (gross basis):

- Baytown refinery and petrochemical complex, located in the state of Texas, Baytown is the second largest petroleum and

---

[252] ExxonMobil 2016 Summary Annual Report, page 40
[253] ExxonMobil 2016 Financial & Operating Review, page 12
[254] Ibid, pages 68 & 69

petrochemical complex in the U.S.[255], with a processing capacity of up to 561MBPD. Baytown refinery is also the largest global manufacturer of base stocks and synthetic lubes[256]. This refinery has been in operations since 1920 when it had an original capacity of 10MBPD.

- Baton Rouge Complex, which began operations in 1909, is located in Baton Rouge. Louisiana. This site is a world-class refinery and petrochemical complex. This complex includes a 503MBPD crude oil capacity refinery; a chemical products terminal for distribution and storage of petrochemicals; one of the largest Chemical plants in the world, with capacities of 6.6 billion pounds of chemicals per year; a plastics plant with capacities to produce 1.2 billion pounds of plastic per year; polyolefins plant, with capacities to produce 2.9 billion pounds of products per year; the Port Allen Lubricants plant, which ships more than 100 million gallons of products annually[257].

- Beaumont Area Complex, located in the state of Texas, is a refinery that can process 363MBPD of crude oil and produce different refined products[258]. The Beaumont complex, which was originally built in 1903, is composed of a refinery, a chemicals plant and a lubricants plant and manufactures a variety of products, from transportation fuels, to basic petrochemicals like ethylene and propylene, to synthetic fluids and lubricants. The refinery recently underwent a 20MBPD expansion of crude oil processing capacity to increase the site's flexibility to process domestic light crude oils[259].

- Billings refinery, located in the state of Montana, began operations in 1949, with a current capacity today of 60MBPD. This refinery primarily uses crude oils from suppliers in Wyoming and Alberta, while refined products are distributed throughout several U.S.

---

[255] XOM Baytown refinery was the largest refinery in the U.S. for many years until the Motiva Port Arthur Refinery expansion was completed in 2012.

[256] http://corporate.exxonmobil.com/en/company/worldwide-operations/locations/united-states/baytown-area-operations/about

[257] http://corporate.exxonmobil.com/en/company/worldwide-operations/locations/united-states/baton-rouge-operations/about

[258] http://corporate.exxonmobil.com/en/company/worldwide-operations/locations/united-states/beaumont-operations/about-us

[259] ExxonMobil 2016 Financial & Operating Review, page 63

states, such as Montana, Wyoming, Utah, Colorado, South Dakota, Washington and Idaho[260].

- Joilet refinery, built in 1972, is one the newest refineries in the U.S. and has a current crude oil processing capacity of 236MBPD. This refinery is configured to process Canadian crude oils delivered by pipeline and can produce a variety of transportation fuels and other petroleum products[261].

- Sarnia complex, located in Ontario, has a crude oil processing capacity of 119MBPD. This complex, operated by Imperial Oil (a subsidiary of ExxonMobil) produces a variety of transportation fuels as well as valuable petrochemical products. ExxonMobil owns a 69.6% share in all Imperial Oil-operated refineries[262].

- Strathcona refinery, located in Alberta, is also operated by Imperial Oil and is one of Canada's largest refineries with a crude oil processing capacity of 191MBPD. This refinery produces transportation fuels, asphalts, base oils and waxes. This refinery has access to *low-cost* to Canadian crude oils and recent logistics improvements have enhanced its processing flexibility[263].

- Nanticoke refinery, located in Ontario, is also operated by Imperial Oil. This refinery produces a variety of petroleum fuels and has a crude oil processing capacity of 113MBPD[264].

European Refineries

In Europe, at the end of 2016, ExxonMobil had a gross crude oil distillation capacity of 1,877MBPD and owned nine refineries in this region[265]:

- Antwerp complex, located in Belgium, has a crude processing capacity of 307MBPD and includes chemicals operations. Antwerp is the largest ExxonMobil refinery in Europe and is fully integrated with the company's chemicals business[266].

- Fos-sur-Mer, located in France, has a crude oil processing capacity of 133MBPD and the company owns 82.9% of this refinery.

---

[260] http://corporate.exxonmobil.com/en/company/worldwide-operations/locations/united-states/billings-operations/about

[261] http://corporate.exxonmobil.com/en/company/worldwide-operations/locations/united-states/joliet-operations/about

[262] ExxonMobil 2016 Form 10-K, page 23

[263] ExxonMobil 2016 Financial & Operating Review, page 63

[264] http://www.imperialoil.ca/en-ca/company/operations/refining-and-supply

[265] ExxonMobil 2016 Financial & Operating Review, page 69

[266] Ibid, page 64

- Gravenchon, located in France, has a crude oil processing capacity of 239MBPD and ExxonMobil has an 82.9% interest.

- Karlsruhe refinery, located in Germany, has a gross crude oil processing capacity of 310MBPD and ExxonMobil owns 25% of this refinery.

- Augusta refinery, located in Italy, has a crude processing capacity of 198MBPD and the company owns 100% of this refinery.

- Trecate refinery, located in Italy, has a crude processing capacity of 132MBPD and the company owns 74.8%.

- Rotterdam refinery, located in the Netherlands, has a crude processing capacity of 191MBPD and is owned in its entirety by ExxonMobil.

- Slagan refinery, located in Norway, has a crude processing capacity of 116MBPD and is 100% owned by ExxonMobil.

- Fawley refinery, located in the United Kingdom, has a crude processing capacity of 261MBPD and is fully owned by the company. Fawley is the largest refinery in the UK, accounts for 20% of the refining capacity and is integrated with chemical plants[267].

Asia-Pacific Refineries

ExxonMobil has an interest in or operates 9 refineries in the Asia-Pacific region with a gross crude atmospheric distillation capacity of 1,107MBPD[268]:

- Altona refinery, located in Australia, has a crude capacity of 80MBPD and is owned 100% by ExxonMobil.

- Fujian refinery, located in China, ExxonMobil owns 25% of this refinery and has a gross crude oil capacity of 268MBPD.

- Jurong/PAC refinery complex is located in Singapore and has a crude oil processing capacity of 592MBPD, is owned 100% by XOM, and has integrated petrochemical operations. This refinery has the largest base oils production capacity in the region as well as producing a range of fuel products and feedstocks[269].

- Sriracha refinery, located in Thailand, has a gross crude oil processing capacity of 167MBPD and XOM owns a 66% interest.

---

[267] Ibid page 64
[268] Ibid, page 69
[269] Ibid, page 65

Middle East

ExxonMobil has interests in one refinery in the Middle East[270], the Yanbu refinery, located in Saudi Arabia, is a 50/50 joint venture with Saudi Aramco. This refinery has a gross capacity of 400MBPD and has been in operations since 1984.

Marketing

ExxonMobil's Marketing operations sell petroleum products worldwide through its *Exxon, Esso,* and *Mobil* brands. At year-end 2016 the company had 20,783 retail sites:

Location	Owned / Leased Sites	Distributors / Resellers Sites	Total Sites	Petroleum Product Sales (MBPD)	Sales per Site (BPD)
United States	0	10,196	10,196	2,250	221
Canada	0	1,792	1,792	491	274
Europe	2,243	3,649	5,892	1,519	258
Asia Pacific	617	855	1,472	741	503
Latin America	5	771	776	82	106
Middle East/Africa	349	306	655	399	609
Worldwide	**3,214**	**17,569**	**20,783**	**5,482**	**264**

ExxonMobil had petroleum product sales of 5,482MBPD of the following products by regions in 2016[271]:

Region	Motor Gasoline, naphthas (MBPD)	Heating Oils, kerosene, diesel Oils (MBPD)	Aviation Fuels (MBPD)	Heavy Fuels (MBPD)	Lubricants, Specialty and other petroleum products (MBPD)	Total (MBPD)
United States	1,338	470	152	55	235	2,250
Canada	260	135	36	16	44	491
Europe	384	774	81	115	165	1,519
Asia Pacific	162	241	88	149	101	741
Latin America	35	33	2	2	10	82
Middle East/Africa	91	119	40	33	116	399
Worldwide	**2,270**	**1,772**	**399**	**370**	**671**	**5,482**

Lubricants

ExxonMobil is the world's largest producer of base oils, which is a major component in manufacturing lubricants. The company's refineries are a strategic source of supply for their lubricants plants around the world, having nine facilities in the Americas, six in Europe, four in the Middle East

---

[270] Ibid
[271] ExxonMobil 2016 Financial & Operating Review, pages 70 & 71

and Arica and five in Asia[272]. Over the past several years, the company has more than doubled sales of high-value synthetic products, particularly of their segment-leading Mobil 1 motor oils.

Chemical

ExxonMobil's Chemical segment is responsible for manufacturing and selling a wide variety of petrochemical products. The Chemical business supplies olefins, polyolefins, aromatics, and a variety of other petrochemicals. The company had earnings of $4.6 billion from its chemical segment in 2016[273]. ExxonMobil is one of the largest Chemical producers in the world, with operations around the globe, with 90% of their manufacturing capacity being integrated with large refineries or natural gas processing plants, providing a unique access to advantaged feedstocks[274]. Moreover, in the U.S., the company has the highest proportion of ethylene production that is produced from *cost-advantaged* ethane, providing a unique feedstock advantage[275].

Half of ExxonMobil's petrochemicals production capacity is located in the U.S. and include the Baytown, Beaumont and Baton Rouge complexes, which provide premium products to the Americas and worldwide.

ExxonMobil, at year-end 2016, had gross yearly production capacities for the following major petrochemicals[276]:

- Ethylene, 9.0 million metric tons
- Polyethylene, 8.6 million metric tons
- Polypropylene, 2.7 million metric tons
- Paraxylene, 3.4 million metric tons

Region	Ethylene	Polyethylene	Polypropylene	Paraxylene
All Other	-	-	-	0.2
Asia Pacific	2.2	2.1	1	1.6
Europe	0.8	1.3	0.3	0.7
Middle East	1.6	1.4	0.2	0
North America	4.4	3.8	1.1	0.9
Worldwide Capacities	**9.0**	**8.6**	**2.6**	**3.4**

---

[272] Ibid, page 65
[273] Ibid, page 74
[274] Ibid, page 76
[275] ExxonMobil 2017 Analyst Meeting, slide 46
[276] ExxonMobil 2016 Financial & Operating Review, page 80

ExxonMobil Chemicals segment had total product sales in 2016 of 24.9 million metric tons per year, of which 19.7 million metric tons were from commodity chemicals and the remaining from specialty chemicals, following is the regional break down[277]:

- Americas region, 10.5 million metric tons
- Europe/Middle East/Africa, 6.6 million metric tons
- Asia Pacific, 8.0 million metric tons

The company produces other petrochemical products such as propylene, butyl, specialty elastomers, adhesive polymers, and a variety of fluids, Oxo alcohols, synthetics and petroleum additives[278].

Overall Company Metrics

ExxonMobil had a market capitalization of approximately $374.4 billion, consolidated revenues of nearly $219 billion, and earnings of $7.8 billion in 2016. At year-end 2016 ExxonMobil's enterprise value was approximately $421 billion and EBITDA was $31 billion, indicating an EBITDA multiple of 14. For the same period, ExxonMobil achieved a return on capital employed (ROCE) of 3%[279] and has close achieved a ROCE close to 15% over the last five years[280]. In 2016, the company's cash return on capital employed (CROCE) was 11% while Return on Equity (ROE) was 4.8%.

ExxonMobil, thanks to its integrated business is the *only* major international oil company that has *positive* Free Cash Flow[281], which has allowed the company to continue to increase its dividend. ExxonMobil has distributed about 50% of cash flow from operations and asset sales to shareholders in the past five year period. Per share dividend has increased on average by 8.8% per year over the last ten years and since the merger of Exxon and Mobil, the company has distributed $370 billion to shareholders[282]. ExxonMobil had a year-end share price of $90.26, distributed dividends of $2.98 per share, resulting in a dividend yield at year-end 2016 of 3.3%. ExxonMobil experienced a total shareholder return of approximately 20% in 2016. The company had earnings per share of $1.88 and a share price $90.26, indicating a price-earnings ratio of 48.

---

[277] Ibid, page 81
[278] ExxonMobil 2016 Financial & Operating Review, page 80
[279] ExxonMobil published ROCE in 2016 was 3.9% versus our 3% using our book's methodology. Main difference is due to how XOM calculates their capital employed versus the definition in this book.
[280] ExxonMobil 2016 Financial & Operating Review, page 4
[281] ExxonMobil 2017 Analyst Meeting, slide 60
[282] Ibid, slide 16

ExxonMobil generated $22.1 billion of cash flow from operations during 2016. In the same period, the company devoted $13.4 billion or 56% of cash flow from operations to dividends while at the same time repurchasing shares for about $977MM. The company had total capital expenditures of $16.1 billion, and with cash flow from operations of $22.1 billion, resulting in Free Cash Flow of approximately $5.9 billion, and including asset sales and other items, had *adjusted* free cash flow of $9.7 billion[283]. ExxonMobil had total debt of $42.8 billion and equity of $176.8 billion, indicating a debt-to-equity ratio of 24%. The company's current ratio at year-end was 0.526 while working capital was *negative* $23 billion. ExxonMobil's total income tax in 2016 was a *negative* $406MM, indicating an effective income tax of approximately 13%. The company incurred interest expense of $453MM in 2016, while capitalized interest was $224MM, indicating total interest cost of $677MM compared to EBIT or operating income of $8.4 billion translating to an interest coverage ratio of 12.

With a total headcount of 71,100 employees (excluding retail station employees), ExxonMobil had adjusted earnings per employee of $139M while cash flow per employee was $311M in 2016.

Refining & Marketing Metrics

ExxonMobil total throughput volumes in 2016 were 4,269MBPD; totaling roughly 1,578MM barrels for the year. Total refined product sales in 2016 were 5,482MBPD, which translates into 2 billion barrels per year. With an average *net* refining capacity of 4,971MBPD[284] and throughput volumes of 4,269MBPD, ExxonMobil achieved a refinery utilization rate of 86%[285] in 2016. ExxonMobil had total clean product sales[286] of 4,042MBPD, which divided by throughput volumes of 4,269MBPD indicates an *estimated* clean product yield of 95%[287]. The company's Nelson Complexity Index for its refineries was 11.5 as of January 2016[288]. ExxonMobil had 5,482MBPD of petroleum product sales, total service stations of 20,783 around the world indicating an average sale per station of 264BPD.

---

[283] Using the book's definition, FCF was $5.9 billion, while using XOM's definition, which includes asset sales and advances, calculates to be $9.7 billion

[284] Represents ExxonMobil's net share of atmospheric distillation capacity

[285] ExxonMobil 2016 Financial & Operating Review, page68

[286] Gasoline and naphtha sales of 2,418MBPD and heating oils, kerosene and diesel of 1,838MBPD, aviation fuels not included.

[287] Please note that ExxonMobil does not report refinery production by product to calculate a clean product yield as per this book. However, refined product sales as a used as a proxy to calculate this metric.

[288] Marathon Petroleum 2016 Profile book, page 4 Industry Comparison

In 2016, total downstream revenues, excluding Chemicals, were $172 billion, which is roughly 78% of the company's total in the same year. Earnings for ExxonMobil's downstream segment totaled approximately $4.2 billion for the year. Using total refined product yearly sales of 2 billion barrels, downstream earnings per barrel and cash per barrel were estimated at $2.10 and $2.86, respectively. Downstream return on capital employed or ROCE in 2016 was 9% while ROCE for Chemicals was 18.6%[289].

Chemical Metrics

The Chemicals segment had earnings of $4.6 billion, of which $1.9 billion were generated in the U.S. while $2.7 billion were earned outside of the U.S[290]. The Chemicals business had prime product sales of 25 million metric tons per year in 2016, indicating earnings per metric ton of $185.

---

[289] ExxonMobil 2016 Financial & Operating Review, page 74
[290] ExxonMobil 2016 Form 10-K, page 98

Tesoro Corp (TSO)

www.tsocorp.com

> *"Perseverance is not a long race; it is many short races one after the other"* –
> *Walter Elliot*

Company Overview

Tesoro Corporation is one of the largest petroleum refining, logistics and marketing companies in the United States. Originally founded as an exploration & production company in 1968, Tesoro began operations of its first refinery in Kenai, Alaska the next year. The company, although originally heavily focused on refining & marketing operations, had small exploration & production operations focused in the U.S. and Bolivia. In December 1999, Tesoro completed the sale of its U.S. and Bolivia upstream operations, receiving at that time, cash proceeds of more than $300MM[291].

As early as 1993, the company had only one refinery, the Kenai refinery which had a capacity of 72MBPD[292] growing to currently a total of 895MBPD of distillation capacity. The company grew through several acquisitions such as the Anacortes refinery, Salt Lake City and Mandan refineries as well through the acquisition of marketing and midstream assets. Keeping on this acquisition trajectory, Tesoro announced on November 17, 2016 that it will acquire Western Refining, an integrated refining & marketing company with three refineries and a distillation capacity of 254MBPD, making Tesoro the 4th largest independent refining & marketing company in the United States[293].

On August 1, 2017 Tesoro Corp and Tesoro Logistics, will change their combined names from Tesoro to Andeavor, after the successful completion of the merger with Western Refining[294]. The company will also use new logos and new ticker symbols for its main company and MLP.

Tesoro Corporation is headquartered in San Antonio, TX, has 6,300 employees and is listed on the New York Stock Exchange under the ticker symbol "TSO".

---

[291] Tesoro Corp 2000 Form10-K , page 3
[292] Tesoro Corp 1994 Form 10-K, page 3
[293] Tesoro to Acquire Western Refining presentation, November 17, 2016, slide 11
[294] http://www.andeavor.com/news/

Refining

Tesoro, as of 2016, operated seven refineries, primarily located in the Western U.S. with a combined crude oil capacity of 895MBPD. After the acquisition of Western Refining, the company will operate 10 refineries with a combined crude oil capacity of 1,157MBPD[295]:

Region	Refinery	Crude Distillation Capacity (MBPD)	Crude Oil Processed (MBPD)	Total Throughput Volumes in 2016 (MBPD)	Total Utilization
Pacific Northwest	Anacortes Refinery	120	105	124	103%
Mid-Continent	Dickinson Refinery	20	17	14	70%
WNR	El Paso Refinery	135	131	140	107%
WNR	Gallup Refinery	25	23	26	103%
Pacific Northwest	Kenai Refinery	72	63	57	79%
California	Los Angeles Refinery	380	330	364	96%
Mid-Continent	Mandan Refinery	74	62	71	96%
California	Martinez Refinery	166	144	143	86%
Mid-Continent	Salt Lake City Refinery	63	53	58	92%
WNR	St. Paul Park Refinery	102	89	92	94%
Total		1,157	1,017	1,089	95%

Anacortes Refinery

The Anacortes refinery is located 70 miles north of Seattle in Washington State. This refinery has a total refining capacity of 120MBPD and primarily supplies gasoline, jet fuel and diesel to markets in Oregon and Washington. This refinery also produces heavy fuel oils and LPGs. This refinery processes primarily crude oils from Bakken (by rail), Alaskan (by tanker) and Canadian (by pipeline). This refinery can process primarily sweet crude oils and also receives crude oils from Asia and the Middle East. The Anacortes refinery's major processing units include atmospheric distillation, vacuum distillation, deasphalting, naphtha reforming, hydrotreating, FCC, butane isomerization and alkylation units[296].

---

[295] TSO Corp 2016 Form 10-K, pages 4-7 & Tesoro to Acquire Western Refining presentation, slide 8
[296] Anacortes Refinery Fact Sheet & Tesoro Corp 2016 form 10K, page 5

Dickinson Refinery

The Dickinson refinery located in North Dakota is the first refinery in the U.S. to be built in over 30 years. This refinery has a capacity of 20MBPD and primarily processed crude oil from North Dakota delivered by pipeline. This refinery primarily produces ultra-low sulfur diesel, naphtha and residual fuel oil, but does not produce motor gasoline or jet fuel since the local demand is short diesel due to extensive drilling in Bakken shale. This refinery was acquired in 2016 from WBI Energy[297].

El Paso Refinery

This refinery was acquired by Tesoro as part of its Western Refining acquisition in 2017. This refinery has a crude oil distillation capacity of 135MBPD and is located in El Paso, Texas. This refinery has historically run a high percentage of light sweet crude oils like WTI and has produced a high percentage of transportation fuels. This refinery also has some flexibility to process sour crude oils from nearby Permian shale production that gives this refinery a unique competitive advantage. This refinery has pipeline access to the Permian Basin through a Kinder Morgan pipeline that provides them access with plentiful crude oils from fields that have long reserve lives[298]. This refinery supplies the Tucson and Phoenix markets in Arizona, as well as Albuquerque, NM, El Paso, TX as well exports refined products to Juarez, Mexico[299].

Gallup Refinery

The Gallup Refinery, also part of Western Refining, is located in Gallup, New Mexico and has a crude oil distillation capacity of 25MBPD. The crude oils processed at this refinery are mainly sourced from regionally produced crude oils from New Mexico, Colorado and Utah while refined products are marketed in the same market areas as well as Arizona.

Kenai Refinery

Located on the Cook Inlet, near Anchorage, Alaska, the Kenai refinery can process up to 75MBPD of crude oil. This refinery's major units include atmospheric distillation, vacuum distillation, hydrocracking, hydrotreating, naphtha reforming and other units, which produce transportation fuels, including gasoline, blend stocks, jet fuel, diesel, fuel and other products like heating oil, heavy fuel oil, LPGs and asphalt.

---

[297] http://www.ogj.com/articles/2016/06/tesoro-acquires-north-dakota-refinery.html
[298] Western Refining 2016 Form 10-K, page 6
[299] Ibid

Los Angeles Refinery

The Los Angeles refinery is the largest on the Western United States and can process up to 380MBPD of crude oil. This refinery processes heavy crude oils from the San Joaquin Valley and Los Angeles Basin, as well as crude oils from Alaska, South America and West Africa[300]. This refinery produces gasoline, jet fuel, diesel, petroleum coke, fuel oil, propylene and calcined coke. This refinery markets its refined products to Southern California, Arizona and Nevada and is set to produce gasoline and diesel with the stringent requirements associated with the California Air Resources Board (CARB). This refinery also has a power cogeneration facility, which is the largest in the state of California.

Mandan Refinery

The Mandan refinery is located in North Dakota has a capacity to process 74MBPD of crude oil and began operations in 1954. This refinery produces gasoline, diesel, jet fuel, heavy fuel oils and LPGs. The Mandan refinery primarily processes domestic light sweet crude oils from North Dakota.

Martinez Refinery

This refinery is located in a city with the same name in California and has a crude distillation capacity of 166MBPD. This refinery processes crude oils from California as well as international crude oils and produces a variety of transportation fuels including CARB gasoline, CARB diesel, conventional gasoline and other products. This refinery has a variety of units, including hydrocracking, delayed coking, naphtha reforming, hydrotreating, FCC and alkylation units.

Salt Lake City Refinery

The Salt Lake City refinery is the largest refinery in Utah and has a crude distillation capacity of 63MPD. This refinery processes crude oils from Utah, Colorado and Wyoming and produces transportation fuels as well as LPG. Refined products from this refinery are marketed through Utah, Idaho, Washington as well as Nevada and Wyoming. This refinery began operations in 1908.

St. Paul Park Refinery

This refinery became part of Tesoro as part of the Western Refining acquisition in 2017 and was originally built in 1939. This refinery can process up to 102MBPD and can process a variety of light, heavy, sweet

---

[300] Tesoro Los Angeles Refinery Fact Sheet

and sour crude oils and converts these into transportation fuels, as well as kerosene, asphalt, propane, LPG and other products.

Marketing

Tesoro's marketing operations sells gasoline and diesel in the Western U.S. through *branded* and *unbranded* channels. Both Tesoro and Western Refining's marketing operations are unique in the fact that their branded stations are branded using a variety of non-company owned brands, such as Exxon, Mobil, Shell, Arco, USA Gasoline, Giant, SuperAmerica as well as the Tesoro brand.

Tesoro's and Western Refining's combined marketing operations will have a total retail site count of 3,015 stations:

Area	Tesoro	Western Refining	Combined Company Total
Pacific Northwest	198	-	198
Mid-Continent	681	285	966
California/Southwest	1,591	260	1,851
Total	2,470	545	3,015

The company's combined marketing operations will be underpinned by several key value drivers[301]:

- Leverages existing brand portfolio for both companies.
- Provides improved *ratable* offtake volumes for their entire refining system, meaning that they will have better integration between both companies refining & marketing operations and ensuring that volumes are contracted or have a guaranteed offtake.
- Mitigates exposure to Renewable Identification Numbers or RINS price volatility.

Overall Company Metrics

Please note that the following metrics are largely for Tesoro Corp and does not include Western Refining since the merger between the two companies was completed after each company filed their own annual reports or 10-K forms with the Securities & Exchange Commission.

At year-end 2016, Tesoro had a market capitalization of $10.2 billion while Western Refining had a market cap of $2.4 billion. Tesoro's consolidated revenues for the same period were $24.6 billion, earnings were $734MM and adjusted earnings were $495MM. Tesoro had Earnings Before Interest

---

[301] Tesoro to Acquire Western Refining presentation, November 17, 2016, slide 12

& Taxes or EBIT of $1.5 billion, EBITDA of $2.3 billion, gross margin of $4.9 billion, indicating a gross margin percentage of 20% and a net margin of 2.9%. Tesoro had cash and cash equivalents of $3.3 billion, total debt of $6.9 billion, capital lease obligations of $53MM and a minority equity value of $2.7 billion, indicating an Enterprise Value or EV of $16.5 billion, translating to an EBITDA over EV ratio of 7.1.

Tesoro's ROCE in 2016 was 5.2% while CROCE was 10.8% for the same period. Tesoro had a return on equity of 4.68% in 2016, based on earnings of $734MM and average equity of $18.4 billion. Tesoro had a beginning share price at 2016 of $105.37, ending share price of $87.45, dividends per share of $2.10 and earnings per share of $6.19, indicating a dividend yield of $2.4, price-earnings ratio of 14 and a total shareholder return of *negative 15%* for 2016. This lower shareholder return or TSR in 2016 is primarily attributable to lower refining margins, causing earnings to go from $1.5 billion in 2015 to $0.7 billion in 2016.

Tesoro generated cash flow from operations of $1.3 billion, using $894MM for capital expenditures, $249MM for dividends, $250MM for share repurchases, which translates to about 38% of CFO being used for shareholder distributions and Tesoro having Free Cash Flow from operations of $410MM. The company had total debt of $6.9 billion, total ending equity of $20.4 billion, indicating a debt to equity ratio of 34%. Tesoro, at year-end 2016, had current assets of $7.4 billion, current liabilities of $3.6 billion, resulting in a current ratio of 2.1 and a *positive* working capital of $3.9 billion. In 2016, Tesoro incurred interest expense of $274MM and had capitalized interest of $31MM, and with EBIT of $1.5 billion, the company's interest coverage ratio was 4.86.

Refining Metrics

At year-end 2016, Tesoro's Refining capacity, based on crude oil distillation, was 895MBPD and the company's refining portfolio had a nelson complexity index of 13.5[302]. In 2016, the company had total throughput volumes of 825MBPD, of which 176MBPD were heavy crude oil volumes, 598MBPD light crude oil and other feedstocks were 51MBPD, resulting in a 92% utilization rate.

In 2016, Tesoro produced 451MBPD of gasoline and gasoline blend stocks (51%), 189MBPD of diesel (22%), 118MBPD of jet fuel (13%) and the remaining 122MBPD (14%) in heavy fuel oils, residual fuel and other

---

[302] Tesoro's Investor Presentation at BAML Refining Conference, March 2nd, 2017 slide 4

products. Tesoro had a gross refining margin of $3,146MM in 2016, total yearly throughput volumes 301,125MBBLs, which translate to a *realized crack spread* of $10.45 per barrel. Refining had operating income of $830MM, D&A expenses of $588MM, indicating that refining earnings and cash per barrel were $1.67 and $2.75 respectively.

Marketing Metrics

In 2016, Tesoro's marketing segment generated $15,490MM in revenues, of which the substantial majority were fuel revenues $15,405MM[303] with the rest being *non-fuel* revenues. The company's gross margin for this segment was $1,198MM in total, of which $1,130MM was fuel and the rest in non-fuel sales. The gross margin percentage as total for the segment was 7.3% for fuel sales while in non-fuel margin was 80%[304] . This segment had EBITDA in 2016 of $889MM and segment operating income of $830MM. The company's marketing operations sold more than 8,879MM gallons of fuel or the equivalent of 580MBPD in 2016. The company's fuel margin per gallon in 2016 was 12.7 cents or $5.33 per barrel for the same period. Out of the company's total station sites of about 2,492, 594 were operated by MSO or *multi-site operators* with the remaining 1,898 being operated by jobber or dealers, which are retail stations owned by a third party that sells products purchased through Tesoro. With 2,492 stations and more than 211MMbbls of refined products sales, on average, each Tesoro affiliated service station had average sales of 232BPD.

---

[303] Tesoro Corp Supplemental Financial and Operational Information, 1Q 2017, page 11
[304] Ibid

Delek US Refining
www.delekus.com

"A goal is a dream with a deadline" – Napoleon Hill

Company Overview

Headquartered in Brentwood, Tennessee, Delek US Holdings is a diversified refining & midstream company with operations in three segments, petroleum refining, midstream logistics and convenience store retailing[305]. Historically, the company has grown through acquisitions, exemplified by its acquisition of the Tyler and El Dorado refineries in 2005 and 2011, respectively[306].

Delek US Holdings is listed on the New York Stock Exchange (NYSE) and trades under the ticker symbol DK since 2006. The company is headquartered in Brentwood, Tennessee and has more than 1,300 employees, of which 650 were employed in the refining segment, 475 in the logistics segment and the remaining in corporate functions[307].

On January 2017, Delek US Holdings announced that it was going to acquire the remaining shares of Alon USA that did not own, creating the country's 7th largest independent refiner of petroleum products in the US, with a crude oil throughput capacity of 302,000 bpd[308].

Combined Company Overview

As previously mentioned, the combined Delek/Alon USA company will have more than 300MBPD of crude capacity, creating a unique Permian-focused refining company with a broadened marketing reach[309]. The company will have assets in the following business segments:

- Refining, with two refineries in Texas, one in Arkansas and one in Louisiana
- Logistics assets, with 20 terminals and 8.5 million barrels of storage capacity.
- Retail with 304 stores and fuel contracts with approximately 640 branded sites.

---

[305] http://www.Delekus.com/aboutus
[306] Delek US Holdings Inc 10-K Form, page 4
[307] Ibid, page 19
[308] Delek US Acquires Remaining Interest in Alon USA IR presentation, January 3, 2017
[309] Ibid, page 3

Refining

The combined Delek & Alon USA will have four operating refineries with a combined crude oil distillation capacity of 302MBPD with an approximate nelson complexity index of 10.3. One of the key differentiators of these two companies is the fact they will become the refining company with the highest percentage of crude slate from the Permian Basin[310]:

Refining Company	Permian Crude Access as % of Crude Slate
Delek/Alon USA	69%
Western Refining	50%
HollyFrontier Corp	30%
CVR Refining	25%
Tesoro/Western Refining Combined	10%
Valero	9%
Phillips 66	4%
Marathon Petroleum	3%
PBF Energy	2%

This proximity and access to Permian crude oils could allow the combined company to capture a $1 per bbl differential, which equates to about $75MM EBITDA[311]. The following table provides an overview of the combined company's active refineries:

Refinery	Crude Oil Distillation Capacity (MBPD)	Throughput volumes (MBPD)	Utilization Rate	Nelson Complexity Index	Clean Product Yield
El Dorado	80	76	96%	10.2	90%
Tyler Refinery	75	72	95%	8.7	94%
Big Spring Refinery	73	71	98%	10.5	86%
Krotz Spring Refinery	74	68	92%	8.4	87%
Total Delek/Alon	**302**	**287**	**95%**	**9.5**	**89%**

El Dorado Refinery

El Dorado refinery with a crude oil distillation capacity of 80MBPD is the largest refinery in Arkansas and represents more than 90% of the refining capacity in that state[312]. This refinery is designed to run a variety of crude oils from light to heavy sour crude oils and receives crude oil by local pipeline and rail cars. This refinery has a nelson complexity index of 10.2 and produces a wide range of refined products, from various grade of

---

[310] Delek US Holdings – Investor Presentation June 2017, slide 9
[311] Ibid
[312] Delek US Holdings 2016 Form 10K, Page 10

gasolines, diesel fuel, LPGs, refinery grade propylene as well as asphalt products. The refined products produced at this refinery are sold to wholesalers and retailers through spot sales, commercial contracts and exchange agreements in several markets. This refinery produced 41MBPD of gasoline, 27MBPD of diesel & jet fuel, 5.2MBPD of asphalt and 2MBPD of other products in 2016[313]. For the same year, El Dorado had a realized crack spread or gross margin per BBL or $2.28, while opex per BBL was $3.73, indicating a *negative* operating margin of $1.45 per BBL or total *loss* of $41.3MM.

El Dorado refinery is equipped with the following process units[314]:

El Dorado Refinery Process Units	Capacity in MBPD except as noted
Crude Oil Atmospheric Distillation unit	80
Crude Oil Vacuum Distillation unit	55
Distillate hydrotreating unit	35
Fluid catalytic cracking unit	20
Naphtha hydrotreating unit	18
LSR naphtha hydrotreating unit	8
Gas oil hydrotreating unit	21
Hydrogen steam methane reforming unit (MCFD)	10
Gasoline hydrotreating unit	9
Continuous catalyst regeneration reforming Unit	15
Isomerization unit	8
Sulfuric alkylation unit (alkylate production capacity)	5

Tyler Refinery

The Tyler refinery has crude oil distillation capacity of 75MBPD and is designed to process mainly light sweet crude oil. Tyler has access to crude oil pipeline systems that allow Delek to access a variety of crude oils, primarily from the Permian basin, East Texas and certain foreign and Gulf of Mexico crude oils. In 2016, this refinery processed 84% WTI crude oil, with the remaining being East Texas crude oil. This refinery has a nelson complexity index of 8.7, which combined with its FCC unit and light crude oils being processed allow this refinery to achieve a 97.8% light product yield (including NGLs). This refinery produces two grades of gasoline, 93 octane and 87 octane as well as aviation gasoline. The distillate fuels produced at this facility are ultra-low sulfur diesel as well as *commercial* and *military grade* jet fuel. In addition to these products, Tyler produces small amounts of propane, refinery grade propylene, butanes, petroleum coke,

[313] Ibid, page 61
[314] Ibid, page 11

sulfur and other blendstocks. In 2016 Tyler produced 39MBPD of gasoline, 28MBPD of diesel & jet fuel, with the remaining being production of NGLs and other products. In 2016, Tyler achieved a realized crack spread or gross margin per BBL of $6.76, while opex per bbl was $3.73, translating to an operating margin per barrel of $3.03 or total for the year of $80.5MM.

The Tyler refinery is equipped with the following process units:

Tyler Refinery Process Units	Capacity in MBPD except as noted
Crude Oil Atmospheric Distillation unit	75
Crude Oil Vacuum Distillation unit	24
Distillate hydrotreating unit	36
Naphtha hydrotreating unit	28
Fluid catalytic cracking unit	20
Continuous catalyst regeneration reforming Unit	18
Gasoline hydrotreating unit	14
Delayed coking unit	8
Sulfuric alkylation unit (alkylate production capacity)	5

Big Spring Refinery

The Big Spring refinery has a crude oil distillation capacity of 73MBPD and is located in the Permian Basin in West Texas. This refinery has a nelson complexity factor of 10.5, which indicates that this refinery has the flexibility to process a variety of crude oils into higher-valued refined products. Big Spring can process significant volumes of sour crude oil and produce a high-percentage of transportation fuels. All of the crude oil processed at this refinery is West Texas crude oil, which is priced at a *discount* on a Midland basis instead of being priced at the delivery point of the WTI, which is Cushing, OK. This refinery produces ultra-low sulfur gasoline, ultra-low sulfur diesel, jet fuel, petrochemicals as well as asphalts. This refinery has the following process units:

Big Spring Refinery Process Units	Capacity in MBPD
Crude Oil Atmospheric Distillation unit	73
Crude Oil Vacuum Distillation unit	24
Fluid Catalytic Cracking unit	25
Catalytic Reforming	21
Diesel & Kerosene Desulfurization unit	28
Naphtha Desulfurization unit	26
Fuel Solvents Deasphalting Unit	10
Heavy Gas Oil Desulfurization unit	7
Alkylation Unit	5
Asphalt	8
Aromatics Extraction Unit	1

The Big Spring refinery is the closest refinery to Midland, TX which allows the company to efficiently source West Texas Sour & West Texas Intermediate at *discounted prices* in comparison to peers, which have to rely on pricing based out of Cushing, OK or out of the Gulf Coast. In addition, this refinery has the ability to source locally-trucked crude oils, which enables the company to control crude oil quality and not have to purchase blended crude oils[315]. Refined products produced at this refinery are primarily marketed in markets such as Central and West Texas, Oklahoma, New Mexico and Arizona, and these products are distributed through a network of pipelines and terminals that the company owns or has long-term agreements. In 2016, the Big Spring refinery had a realized crack spread or margin of $8.28 per barrel, while opex per barrel was $3.73, indicating an operating margin per barrel of $4.55 or $119MM per year[316].

Krotz Springs Refinery

The Krotz Springs refinery is located in Central Louisiana and has a crude oil distillation capacity of 74MBPD and a Nelson complexity index of 8.4. This refinery is centrally located to access crude oils via barge, pipeline, railcar and truck. On the refined products side, this refinery has direct access to the Colonial Pipeline, one of the largest refined products distribution systems in the United States. This refinery processes primarily light sweet crude oils and can typically convert up to 90% of its feedstock into higher-valued finished products, such as gasoline, diesel, jet fuel and others. This refinery does receive crude oils from the Permian basin via a combination of pipelines and barges, although in recent years since differentials have narrowed, this refinery has shifted to relying less on crude oils from the Permian basin. The Krotz Springs refinery has the following process units:

Krotz Springs Refinery Process Units	Capacity in MBPD
Crude Oil Atmospheric Distillation unit	74
Crude Oil Vacuum Distillation unit	36
Fluid Catalytic Cracking unit	34
Catalytic Reforming	13
Gasoline Desulfurization unit	18
Naphtha Desulfurization unit	14
Isomerization unit	6

---

[315] Many blends of crude oils create significant challenges for refineries because of the "mixed properties" of these crude oils. Some of these blends even create process upsets at the refineries. See more: http://www.tandfonline.com/doi/abs/10.1081/LFT-120018539?scroll=top&needAccess=true&journalCode=lpet20

[316] Alon USA 2016 Form 10-K, pages 3-5 and pages 40-41

This refinery markets its products through the Colonial pipeline system, which then feeds into wholesale markets in the Southern and Eastern United States. Other products, such as propylene/propane mix are sold via railcar and truck to consumers at Mont Belvieu in Texas or in markets in Louisiana[317]. In 2016, the Krotz Springs refinery's realized crack spread was $3.06 per BBL, while opex per bbl was $3.78, indicating an operating *loss* per barrel of $0.72 or *losses* of $18MM per year.

Logistics, Marketing & Retail

Delek currently owns nine light product distribution terminals in Nashville, Memphis, Tyler, Big Sandy, San Angelo, Abilene, Mount Pleasant, North Little Rock and El Dorado. These terminals provide wholesale marketing customers with terminaling and storage services to market refined products out of the company's refineries. The following table provides an overview of the logistic assets and throughput volumes in 2016:

Logistic Assets	Average Volumes (BPD)
West Texas marketing throughputs	13,257
Terminalling throughputs	122,350
East Texas marketing throughputs	68,131
Lion Pipeline System: Crude pipelines (non-gathered)	56,555
Lion Pipeline System: Refined products pipelines to Enterprise Systems	52,071
SALA Gathering System	17,756
East Texas Crude Logistics System	12,735
Total Throughput Volumes	**342,855**

The merged Delek Alon USA company will own two master limited partnerships, Alon USA Partners (NYSE: ALDW) and Delek Logistics Partners (NYSE: DKL).

The combined company will own substantial asphalt operations with a market presence from Tennessee to California, have combined sales of almost 1 million tons per year and have the following assets[318]:

- 15 Asphalt Terminals
- 76 Asphalt Trailers
- 650 Railcars

---

[317] Alon USA 2016 Form 10-K, page 4
[318] Delek US Holdings Inc. to Acquire Remaining Shares of Alon USA, January 3, 2017, slide 14

Alon USA's Asphalt operating segment had the following key indicators in 2016:

- Blended asphalt sales volume of 522,000 tons & non-blended sales volumes of 85,000 tons.
- Asphalt sales price per ton of $399 while non-blended price of $146.
- Asphalt margins of $99 per ton and capital expenditures of $3MM.

Alon USA will bring in into the merged company retail operations that are substantial in nature. A few of the key statistics from this segment in 2016[319]:

- 306 retail store system.
- Retail fuel sales of 13,631BPD and average fuel sales of 45BPD per retail station.
- Retail fuel margins of $8.32 per gallon.
- Total sales of $731MM per year, of which $324MM or 44% are non-fuel or merchandise sales.
- Merchandise sale margin of 31.2% or $101MM per year.
- Wholesale operations that supply about 640 branded sites, including the 306 company owned or leased sites.
- Wholesale fuel sales volumes of more than 1 billion gallons per year.
- The largest 7-Eleven licensee in the United States.

Overall Company Metrics

The following company and refining metrics have been calculated on a *pro-forma* basis, which assumes that both entities were emerged for the entire year 2016. Certain pro-forma items have been cross-checked against Delek's SEC form S-4 Registration Statement.

In 2016, Delek Refining had consolidated revenues of $8.1 billion, cost of goods sold of $7.2 billion, gross margin of $932MM, gross margin percentage of 11%, operating *losses* of $52MM, adjusted EBITDA[320] of $218MM, GAAP *losses* of $97MM[321] and adjusted *losses* of $24MM. Had Alon USA and Delek been combined, this company would have had

---

[319] Alon USA 2016 Form 10-K, page 43
[320] Adjusted for non-cash items
[321] Delek Holdco Inc. Form S-4 Registration Statement, page 223

81,172,580 shares[322] and an estimated share price of $28.42 at year-end 2016, earnings per share of *negative* $1.20 and dividends per share of $0.99, indicating a hypothetical market cap of $2.3 billion, price earnings ratio of *negative* 24 and a dividend yield of 3.48%. With a hypothetical beginning share price of $31.89, the combined company would have provided shareholders with a total shareholder return of *negative* 7.8% in 2016, primarily due to narrowing crude differentials and lower crack spreads in 2016 than in prior years. With total debt of $1.4 billion, minority interest value of $252MM, market cap of $2.3 billion and cash & cash equivalents of $826MM, this company would have had an Enterprise Value of $3.1 billion, translating to an EV/EBITDA multiple of 14.

With adjusted losses of *negative* $24.5MM, after-tax interest expense of $55.2MM, D&A expenses of $197.5MM[323], average capital employed of $3.9 billion, the combined company would have had ROCE of 1.39% and CROCE of 6.5%. Return on equity in 2016 would have been a *negative* 3.90%.

The combined company would have generated $328MM in cash flow from operations, incurred $135MM in capital expenditures, indicating that free cash flow would have been $193MM in 2016. The company would have distributed $80MM in dividends to shareholders and repurchased $6MM in shares, translating to a 26% of cash flow from operations being devoted to shareholder distributions. The company's total debt would have been $1.4 billion, total ending equity would have been $1.8 billion, resulting in a total debt to equity ratio of 77% in 2016. The company's *pro forma* current assets and current liabilities would have been $1.9 billion and $1.4 billion, respectively, translating to a current ratio of 1.362 and working capital of *positive* $502MM. With interest expense of $124MM and an adjusted EBIT of $21MM, the interest coverage ratio would have been 0.16 in 2016.

Refining Metrics

Delek and Alon USA's combined refining capacity is 302MBPD, while throughput volumes were 287MBPD, indicating a utilization rate of 95%. The combined company processed a total of 271MBPD of crude oil in 2016, with the majority of these crude oils originating from the Permian basin. These two companies produced combined gasoline volumes of 148MBPD, diesel and jet fuel volumes of 107MBPD, while total refined product volumes were 286MBPD, translating to a clean product yield of

---

[322] Ibid
[323] Ibid

89%, one of the highest in the industry. The company's combined nelson complexity index is 9.5. With a gross margin of $932MM in 2016 and total throughput volumes of 105MMBBL, the combined company's realized crack spread per barrel was $8.89. With total yearly refined products production of 104.5MMBBL, adjusted losses of $24.5MM, D&A expenses of $197.5MM, adjusted earnings per barrel were *negative* $0.23 and cash per barrel was $1.66 in 2016. Operating expenses were $443MM, translating to opex per barrel of $4.22.

Marketing Metrics

Alon USA has substantial marketing operations, which will be incorporated into Delek US. The company had total fuel sales of 208,963 thousand gallons per year or the equivalent of 13,631BPD in fuel sales. With the company's retail station portfolio being 306 stores, the average fuel sale per station was 45BPD in 2016. The company's average retail fuel margin was 19.8 cents per gallon or $8.32 per BBL. The average retail fuel sales price was $1.95 per gallon. Total sales for the retail segment were $732MM, of which $324MM or 44% of all sales were merchandise sales and the rest being fuel sales. Merchandise sales had a gross margin of 31.2% or $101MM per year out of a total gross margin of $143MM. In other words, out of the retail segment's gross margin, 70% of that margin was made up by merchandise sales with the rest being fuel.

Chapter VI – Europe, Russia and CIS

"Our greatest weakness lies in giving up. The most certain way to succeed is always to try just one more time." – Thomas Edison

Rosneft (OJSCY)

www.rosneft.com

"Either I will find a way, or I will make one" – Philip Sidney

Company Overview

Rosneft is the world's largest *publicly traded* National Oil Company in terms of proved oil reserves and liquid hydrocarbons production[324]. Moreover, Rosneft is also Russia's largest oil producer and the third largest natural gas producer in the country. The company is also Russia's leading oil refiner and the country's largest taxpayer[325]. . The Russian Federation, through its OJSC Rosneftgaz Company, owns 50% of the company, BP owns 19.75%, a Swiss-Qatar consortium owns 19.50% and the remaining 10.75% of the shares are publicly traded on the Moscow Stock Exchange and the London Stock Exchange. The company is headquartered in Moscow and had a total employee headcount of about 295,000 people at the end of 2016.

Rosneft predecessor companies' history dates back to the 19th century, when Russian entrepreneurs started oil-field exploration in Sakhalin, a large island in Eastern Russia[326]. The modern day Rosneft was established in 1993 when more than 250 industrial entities were placed into a trust of Rosneft. In 2013, Rosneft acquired a 100% stake in the TNK-BP joint venture with BP receiving $12.5 billion in cash plus 18.5% of Rosneft shares, which in addition to BP's prior 1.25% ownership resulted in the current 19.75% interest that BP has in Rosneft[327]. On December 2016, the Russian government announced that it would be selling 19.75% of its shares in Rosneft to the Qatar Investment Authority, and Glencore, a leading commodities producer and trader[328]. Today, Rosneft has a vast portfolio of oil & gas assets, and with its strategic partners, ExxonMobil and Statoil, is undertaking unprecedented exploration & development efforts in the Artic[329].

In July 2006, Rosneft completed an IPO, offering its shares on both the London and Moscow Exchanges. Rosneft trades on the Moscow Stock Exchange under the symbol ROSN and on the London and New York

---

[324] Rosneft 2016 Annual Report, page 6
[325] Ibid
[326] www.rosneft.com/about/history/
[327] http://www.bp.com/en/global/corporate/about-bp/bp-worldwide/bp-in-russia.html
[328] https://www.rosneft.com/press/releases/item/185049/
[329] http://www.rosneft.com/about/history/

Stock Exchanges as a Global Depositary Receipts (similar to ADRs) under the symbols ROSN and OJSCY.

Please note that the metric ton to barrel conversion factor used was 0.1364 metric ton to one 1 barrel, which may differ depending on the *actual gravity* used by Rosneft.

Areas of Operation

Rosneft is an integrated oil & gas company with operations in:

- Exploration & Production of hydrocarbons, primarily inside the Russian Federation.
- Refining & Marketing of hydrocarbons.
- Transportation of hydrocarbons, refined products, natural gas and other products.
- Petrochemicals.
- Natural Gas Processing.

Refining

Rosneft is Russia's largest refining company and currently has a total refining capacity of more than 2,400MBPD. With the acquisition of Bashneft, three new refineries were included in Rosneft's operations. For several years Rosneft has consistently implemented a refinery upgrade program, which has helped improved the quality of Rosneft's refined petroleum products. As part of this modernization program, Rosneft completed a transition to 100% production of Euro 5[330] gasoline for the Russian domestic market.

The company produced more than 2,040MBPD of refined petroleum products and petrochemicals in 2016[331]:

Product	2016 Production in MBPD
Diesel fuel	674
Motor gasoline	362
Fuel oil	497
Naphtha	141
Kerosene	67
Petrochemicals	30
Others	271
Total in Russia	**2,043**

---

[330] Euro 5 is a specification for motor gasoline regulated by the European Union which sets very stringent limits on certain components.
[331] Rosneft's Annual Report 2016 page 112

Refining Overview

At the end of 2016, Rosneft had a portfolio of 13 large refineries in Russia and abroad with the following estimated capacities, throughput, clean products yield and conversion rates[332]:

Refinery	Estimated Design Capacity (MBPD)	Refinery Throughput (MBPD)	Utilization Rate (%)	Clean Products Yield (%)	Conversion Rate (%)
Bashneft Refineries	466	368	79%	67%	86%
Ryazan	378	307	81%	57%	70%
German Refineries	251	255	102%	79%	94%
Tuapse	241	197	82%	49%	66%
Angarsk	205	185	90%	64%	77%
Novokuibyshev	177	143	81%	55%	75%
Syzran	171	127	74%	59%	73%
Komsomolsk	161	125	78%	59%	80%
Achinsk	151	143	95%	56%	70%
Saratov	141	119	84%	51%	80%
Kuibyshev	137	123	90%	54%	60%
Total	**2,477**	**2,089**	**72%**	**57%**	**86%**

Bashneft's refineries

Rosneft acquired Bashneft's assets in 2016, and these include a refinery in Ufa, Russia, which is the largest in the country with an estimate crude oil distillation capacity of 466MBPD. This refinery is actually several refineries, including Ufaneftekhim, Novoil and UNPZ, forming part of this integrated refinery complex in the city of Ufa. This complex also includes a petrochemical plant and a gas processing plant. Rosneft expects several improvements to occur such as changing the composition of the feedstock and optimization of the maintenance program[333].

Ryazan Refinery

Ryazan refinery is the *largest single* refinery in Russia with a crude oil distillation capacity of 378MBPD, conversion rate of 69.8% and clean product yield of 56.6%. This refinery began operations in 1960 and currently this refinery produces a wide range of high quality products, including Euro 5 diesel fuel, jet fuel, finished motor gasolines and other products.

---

[332] Rosneft's Annual Report 2016 pages 110-119
[333] Ibid, page 113

German Refineries

Rosneft's German refineries include the MiRO refinery, PCK Raffinerie and Bayernoil refinery. As a result of the restructuring of the joint venture Ruhr Oel GmbH (ROG) completed in 2016, Rosneft gained control of over 12% of the refining capacity in Germany with a total throughput of 251MBPD (net to Rosneft). The company became the third largest refining company in the German market and started developing its own business in the country. The following are the three refineries in Germany[334]:

- MiRO Refinery, with an approximate total crude distillation capacity of 300MBPD, is located in Karlsruhe in the German state of Baden-Württemberg. MiRO is Germany's largest and one of Europe's most advanced refineries. Rosneft owns a 24% interest and the nelson complexity factor is 9.4.

- PCK Refinery is located in Schwedt in the German state of Brandenburg, east of Berlin, has a total crude distillation capacity of 233MBPD and a complexity index of 9.8. Rosneft owns a 54.17% interest in this refinery and is supplied with Urals oil through the Druzhba pipeline.

- Bayernoil Refinery is located in Neustadt der Donau in the German state of Bavaria, has a total crude oil distillation capacity of 207MBPD and a nelson complexity factor of 6.8. Rosneft owns 25% of this refinery and Bayernoil refinery supplies fuel to Bavaria and northern Austria.

Tuapse

Tuapse is Rosneft's oldest refinery and was commissioned in 1929 and it is the only Russian refinery in the Black Sea region. Tuapse processes crude oil from Western Siberia, which is delivered via pipeline as well as railcars. The refinery's close proximity to the Tuapse loading terminal enables the plant to export 90% of all petroleum products at the refinery[335]. Tuapse has a crude oil distillation capacity of 241MBPD, and has a conversion rate of 66.4%.

Angarsk Petrochemical Company

Angarsk Refinery is located in the southern part of Eastern Siberia and ranks among Russia's largest refineries. This refinery processes West Siberian crude oil delivered by the Transneft pipeline system and is a key

---

[334] Ibid, page 123
[335] http://www.rosneft.com/Downstream/refining/Refineries/Tuapse_Refinery/

refined products supplier in Siberia and the Russian Far East regions. Angarsk refinery was originally built in 1955 and was acquired by Rosneft in 2007[336].

Novokuibyshev

Novokuibyshev Refinery began operations in 1951 and was acquired by Rosneft in 2007. The refinery processes West Siberian crudes as well as feedstocks from the nearby company's operations in the Samara region[337].

Syzran

Syzran, similar to Novokuibyshev, was built in 1942; this refinery was also acquired by Rosneft in 2007. Syzran also processes West Siberian and other crude oils and produces a wide range of refined products, primarily transportation fuels[338].

Komsomolsk

Komsomolsk Refinery: this refinery was completed in 1942 and has been part of the company since Rosneft was established. The refinery is located is located in the Khabarovsk region close to the Chinese border in Eastern Russia. The refinery processes West Siberian crude as well as crude oils coming from the Sakhalin fields. The refinery's finished products are exported to Japan, South Korea and Vietnam via a marine terminal.[339]

Achinsk

Achinsk Refinery: is the only major refinery in the Krasnoyarsk Territory, central Russia and is a key refined products supplier to neighboring areas. The refinery was acquired by Rosneft in 2007, but was originally built in 1972, with subsequent expansions later on[340].

Saratov

Saratov located in the Russian Federal district of Volga; this refinery was originally built in 1934 and was acquired by Rosneft in 2013 as part of the TNK-BP acquisition. Refined products are delivered to both domestic and export markets.[341]

---

[336] http://www.rosneft.com/Downstream/refining/Refineries/Angarsk_Refinery/
[337] http://www.rosneft.com/Downstream/refining/Refineries/Novokuibyshev_Refinery/
[338] http://www.rosneft.com/Downstream/refining/Refineries/Syzran_Refinery/
[339] http://www.rosneft.com/Downstream/refining/Refineries/Komsomolsk_Refinery/
[340] http://www.rosneft.com/Downstream/refining/Refineries/Achinsk_Refinery/
[341] http://www.rosneft.com/news/news_in_press/250420143.html

Kuibyshev

The Kuibyshev refinery began operations in 1945 and has been part of Rosneft since 2007. This refinery has a crude oil distillation capacity of 137MBPD and has had several upgrades throughout the years, including a vacuum gas oil hydrotreater, hydrogen and sulfur production units, a new fluid catalytic cracking unit and isomerization unit.

Mini-Refineries

Rosneft owns 4 *mini-refineries[342]* in Western Siberia, Eastern Siberia, Timan-Pechora and the southern part of Russian with an overall estimated crude distillation capacity of 116MBPD. The company also has an interest in refining processing agreements and other smaller refineries.[343]

Marketing

Rosneft has the largest retail network in Russia with more than 2,800 stations in Russia and 65 stations in Belarus, Abkhazia and Kyrgyzstan. Rosneft's retail network covered 66 regions in Russia. Rosneft has a 23% market share in the refined petroleum product retail market.

The following table provides a breakdown of Rosneft's retail sales by products on their 2,962 stations, indicating average sales per station of 243BPD:

Product	2016 Domestic Sales in MBPD
Diesel fuel	219
Motor gasoline	293
Fuel oil	41
` Kerosene	65
Bunker Fuel	16
Other	86
Total Sales	**720**

Overall Rosneft sells refined petroleum products in both the domestic Russian market as well as internationally and had total sales of 2,020MBPD in 2016:

- Sales abroad 1,335MBPD.
- Wholesale distribution in Russia 459MBPD.
- Retail sales in Russia 226MBPD.

---

[342] A mini-refinery is a small refinery that fractionates less than 5MBPD of crude oil by atmospheric distillation. Please see here for more information: http://reftexas.com/media/52154b7d16905.pdf
[343] http://www.rosneft.com/Downstream/refining/

Marine Terminals

Rosneft owns three main marine terminals that help the company supply refined petroleum products to international markets:

- Rn-Tuapsenefteprodukt, which provides products to Europe, Asia and the Americas has an estimated capacity of 356MBPD and had throughput volumes of 342 in MBPD. This terminal's products are supplied from the Tuapse, Achinsk, Kuibyshev and Novokuibyshev refineries. This terminal also stores and ships crude oil for the Tuapse refinery.

- Rn-Nakhodkaprodukt Llc is supplied by the Komsomolsk, Achinsk and Angarsk refineries and has a capacity of 157MBPD and throughput volumes of 134MBPD in 2016.

- Rn-Arkhangelsknefteprodukt Llc is supplied with refined products from the Yanos and Ryazan refineries, has a capacity of 84MBPD and only had throughput volumes of 20MBPD in 2016.

Petrochemicals

Rosneft has a substantial presence in the petrochemicals business, selling more than 3.5 million metric tons of petrochemicals in 2016[344]. In addition to selling petrochemicals, Rosneft produces petrochemicals at several sites[345]:

- Ansgark Polymer Plant, this is the largest petrochemical facility in Rosneft's portfolio, which produces ethylene, high-density polyethylene, propylene, benzene, styrene, polystyrene and other products. This facility produces in a year 200,000 MT of ethylene, 100,000 MT of propylene and 60,000 MT of benzene.

- Novokuibyshevsk is one of the largest petrochemical facilities in Russia producing a variety of chemical products such as tert-amyl methyl ether, phenol, ethyl alcohol, acetone, phenolic resin and para-tertiary butylphenol. This facility produced in 2016 1.1MM metric tons of petrochemical products.

- Ufaorgsintez, this facility specializes in the production of phenol, acetone, high-density polyethylene, polypropylene, synthetic rubber, as well as bisphenol A, phenol formaldehyde resins, alkyphenol, plastic films and many other products.

---

[344] Rosneft 2016 Annual Report, page 60
[345] Ibid, page 119

Overall company metrics

Please note that Rosneft's financial statements are *largely* published in Russian Ruble and the exchange rate used to translate those financial statements to U.S. dollars used in this book was 67 RUB per 1 US$, which was the effective exchange rate at December 31, 2016.

Rosneft had 10.6 billion shares outstanding at year-end 2016, and had a Global Depositary Receipt or GDR[346] share price of $6.50, translating into a market cap of $69 billion at the end of 2016. Rosneft, for the same period, had total consolidated revenues of $73 billion, EBITDA of $17 billion, earnings of $2.8 billion and adjusted earnings of $6.7 billion[347]. With an average capital employed of $120 billion and a ROCE numerator of $4.9 billion, Rosneft achieved a Return on Capital Employed or ROCE of 4.1% in 2016[348]. Meanwhile, Cash Return on Capital Employed or CROCE was 10.1%. Rosneft had total average equity of $99 billion, indicating a Return on Equity of 6%. At year-end 2016, Rosneft had cash & cash equivalents of $11.8 billion, capital lease obligations of $657MM, minority interest equity value of $6.2 billion and total debt of $53.5 billion, indicating an Enterprise Value of $117.5 billion, which equates to an EBITDA multiple of 6.9.

With a year-end share price of $6.50, dividends of $0.09 per share, earnings per share of $0.26, Rosneft had an effective dividend yield of 1.37% and a price earnings ratio of 25 at year-end 2016. Rosneft, at the beginning of 2016, had a share price of $3.48, ending share price of $7.62 and dividends of $0.09, resulting in Total Shareholder Return or TSR of 90%, primarily attributable to favorable conditions in the Russian stock market as well as a recovery of the price of oil in 2016. The company's effective income tax rate for the same period was 37%[349]. Rosneft incurred $2.9 billion in interest in 2016 while EBIT was $9.9 billion, translating to an interest coverage ratio of 3.43.

In 2016, Rosneft generated $9.4 billion in cash flow from operations[350], using about $1.9 billion or 20% for shareholder dividends in the same period. Rosneft's capital expenditures in 2016 were $10.7 billion[351] indicating that free cash flow was a *negative* $1.3 billion. Rosneft's total debt at year-end 2016 was $53.5 billion while the company's total equity was

---

[346] GDRs are traded on the London Stock Exchange and are traded in US dollars.
[347] Rosneft Consolidated Financial Statements for the year ended Dec 31, 2016, pages 8-13
[348] Ibid
[349] Ibid
[350] Ibid, page 41
[351] Ibid, page 3

$43.7 billion, translating into a debt-to-equity ratio of 122%. Rosneft had total current assets of $34.3 billion, total current liabilities of $41.4 billion, translating to a *negative* working capital of $7.1 billion and a current ratio of 0.829.

With a total employee headcount of 295,800, adjusted earnings of $6.7 billion and cash flow from operations of $9.4 billion, average adjusted earnings per employee were $23M while average operating cash flows per employee were $32M in 2016.

Downstream metrics

Rosneft had an estimated global refining capacity of 2,370MBPD[352]. In 2016, throughput volumes were 2,034MBPD, indicating a utilization rate of 86%. In 2016, Rosneft had an estimated clean product yield of 58% based on gasoline production of 503MBPD, diesel production of 674MBPD and kerosene production of 1MBPD based on total throughput volumes of 2,034MBPD in 2016. Overall downstream revenues were an estimated $75 billion (including intercompany revenues), adjusted earnings $1.9 billion and D&A $1.3 billion in 2016. Downstream adjusted earnings per barrel were $2.54 while downstream cash per barrel was $4.22 for the same period.

---

[352] Based on design capacity and using a conversion factor of 0.1364 metric ton to 1 barrel

Total S.A. (TOT)

www.total.com

"The more things you do, the more you can do" – Lucille Ball

Company Overview

Total S.A. is one of the world's largest integrated oil & gas companies. Total operates in over 150 different countries, operates in every sector of the oil industry, and has over 102,000 employees[353] at the end of 2016. As an integrated oil & gas company, Total has operations in upstream, midstream and downstream. Total is headquartered in the La Defense district in Courbevoie, France[354].

Total is a multinational, integrated oil and gas company, founded March 28th, 1924. Originally called Compagnie Française des Pétroles (CFP), it was renamed in 1985 to Total CFP; then the CFP was dropped in 1991, and the name further changed to Total FINA after the acquisition of Petrofina of Belgium in 1999. One year later with acquisition of Elf Aquitaine the name was changed to TotalFinaElf; but, finally simplicity reigned over confusion and corporate resolve rebranded the company just Total on May 2003.

Total is listed under the symbol FP on the EURONEXT, TOT symbol on the New York Stock Exchange (NYSE) and TTE on the London Stock Exchange (LSE).

Total produced 2,452MBOED of oil and gas during 2016[355]. By the end of 2016, the company also had nearly 11.5 billion BOE in proved reserves of oil and gas. Total had 102,168 employees as of the end of 2016 of which approximately 31% were from France, 25% of all employees were from the rest of Europe and 44% were from outside of Europe[356].

Areas of Operation

As an integrated company, Total has operations in upstream, midstream, and downstream and is organized into the following 3 reportable operating segments:

- Upstream, which is engaged in oil and gas exploration, liquefied natural gas and power generation

---

[353] TOTAL 2016 Factbook, page 24
[354] TOTAL 2016 Form 20-F, cover page
[355] TOTAL 2016 Factbook, page 27
[356] Ibid, page 26

- Refining & Chemicals, which includes refining, petrochemicals, specialty chemicals, oil trading and shipping activities
- Marketing & Services, which participates in worldwide supply and marketing activities primarily of petroleum products.

Similar to other integrated companies, Total's midstream operations are largely reported as either upstream or downstream.

Refining & Chemicals

Total's Refining & Chemicals segment refines and processes hydrocarbons to produce fuels, lubricants and petrochemicals. Total is among the 10 largest refiners in the world. Total had a net refining capacity of 2,011MBPD at year-end 2016. Total's petrochemical operations are integrated with its refining operations. Total's specialty chemicals businesses include elastomer processing (Hutchinson), adhesives (Bostik) and electroplating chemistry (Atotech).

Refining

In 2016, Total had refining capacity of around 3,187MBPD based on gross crude distillation capacity in 2016 while on a net basis it was 2,011MBPD. For the same period, 682MBPD or 34% of the capacity is located in France, and 772MBPD or 38% is located in other locations throughout Europe.

The remaining capacity is located in the U.S., Africa, the Middle East and Asia[357]:

Region	Refining Capacity MBPD	Throughput Volumes MBPD
France	682	669
Rest of Europe	772	802
United States	202	193
Asia & Middle East	303	249
Africa	52	53
Total	**2,011**	**1965**

---

[357] TOTAL Fact Book 2013, pages 115-116

Total produced a total of 1,871MBPD of refined products in 2016:

Product	Volumes in MBPD
Diesel fuel and Heating Oils	795
Motor gasoline	324
Aviation gasoline, jet fuel and kerosene	182
Fuel Oils	140
LPG	60
Lubricants, Bitumen and other	370
Total Refined Products Sales	**1,871**

France Refining
In France, Total fully owns and operates four refineries with a crude distillation capacity of 682MBPD[358]:

- Gonfreville refinery, located in Normandy, has a refining capacity of 253MBPD.
- Donges refinery has a refining capacity of 219MBPD.
- Feyzin refinery has a refining capacity of 109MBPD.
- Grandpuits refinery has a refining capacity of 101MBPD.

Other European Refining
In the remainder of Europe, Total owns an interest in five refineries with a *gross* crude distillation capacity of 953MBPD:

- Immingham/Lindsey refinery, located in the United Kingdom, has a refining capacity of 109MBPD and is fully owned and operated by Total.
- Vlissingen refinery, located in the Netherlands, has a gross refining capacity of 148MBPD, and Total owns a 55% interest.
- Antwerp refinery, located in Belgium, has a refining capacity of 338MBPD, and is fully owned and operated by Total.
- Leuna refinery, located in Germany, has a refining capacity of 227MBPD, and is fully owned and operated by Total.
- Trecate refinery, located in Italy, has a *gross* refining capacity of 131MBPD, and Total owns a 13% interest.

---

[358] TOTAL 2016 Factbook, page 101

African Refining

In Africa, Total owns an interest in four refineries with a *gross* crude distillation capacity of 252MBPD:

- Limbe refinery, located in Cameroon, has a refining capacity of 45MBPD and Total owns a 20% interest.
- Abidjan refinery, located in Côte d'Ivoire, has a refining capacity of 76MBPD and Total owns a 25% interest.
- Dakar refinery, located in Senegal, has a refining capacity of 24MBPD and Total owns a 20% interest.
- Sasolburg refinery, located in South Africa, has a refining capacity of 110MBPD and Total owns an 18% interest.

Asia & Middle East Refining

In Asia & Middle East, Total owns an interest in four refineries with a *gross* crude distillation capacity of 1,062MBPD:

- Dalian refinery, located in China, has a refining capacity of 219MBPD and Total owns a 22% interest.
- Daesan Refinery, located in South Korea, has a refining capacity of 158MBPD and Total owns a 50% interest.
- Ras Laffan refinery, located in Qatar, has a refining capacity of 300MBPD and Total owns a 10% interest.
- Jubail refinery, located in Saudi Arabia, has a refining capacity of 386MBPD and Total owns a 38% interest.

United States Refining

In United States, Total owns an interest in two refineries with a *gross* crude distillation capacity of 238MBPD:

- Port Arthur refinery, located in Texas, has a refining capacity of 178MBPD and it is fully owned and operated by Total.
- In the same vicinity, Total owns a condensate splitter that a gross capacity of 60MBPD and the company has a 40% interest.

Chemicals

Total is also a large-scale chemicals manufacturer, with petrochemical operations integrated with its refining operations. Total had the following chemical production capacities at year-end 2016, in thousands of metric tons per year[359]:

Product Group	Europe	North America	Asia and Middle East	World
Olefins	4,373	1,525	1,571	7,468
Aromatics	2,903	1,512	2,429	6,844
Polyethylene	1,120	445	773	2,338
Polypropylene	1,350	1,200	400	2,950
Polystyrene	637	700	408	1,745
Others	0	0	63	63
Totals	**10,383**	**5,382**	**5,643**	**21,407**

Total sold petrochemical products around the world and in 2016 had the following sales by area[360]:

Region	Sales percentage
France	12%
Rest of Europe	41%
North America	30%
Rest of world	16%
Total	**100%**

Trading

Total is one of the world's largest traders of crude oil and petroleum products, engaged in both the physical and derivatives markets, both organized and over the counter[361].

Marketing & Services

Total Marketing & Services business segment encompasses three main business areas:

- Retail: with a network of more than 16,000 service stations and a presence in more than 150 countries, these stations provide fuel as well as merchandise and car wash and servicing.
- Lubricants: a high margin business that accounts for more than 1/3 of this segment's earnings that produces and sells lubricants around the world.

---

[359] TOTAL 2016 Factbook, page 103
[360] Ibid, page 103
[361] TOTAL 2016 Form 20-F, page 34 & 35

- Business-to-Business: which focus on delivering innovative solutions in the energy markets, such as hybrid solution (gas & solar), bitumen and bunkering fuel activities centered on offering LNG for ships.

Retail

Total's marketing & retail business encompasses 16,461 service stations that distribute company and third party petroleum products throughout the world with 55% of stations located in Europe, 25% located in Africa, 11% in the Asia/Pacific region, with the remaining sites located in the Middle East and the U.S[362]. The Marketing and Services Segment generated $1.6 billion in adjusted net operating income in 2016, and employed more than 32,000 people[363].

Total's Marketing & Services Segment had consolidated product sales of 3,418MBPD[364], of which 1,749MPBD[365] were attributable entirely to this segment and the remaining were attributable to refining. Total is among the largest fuel marketers in Western Europe and the number one in Africa[366]. This segment is also engaged in supply and trading of refined petroleum products around the world. Total had the following petroleum sales and service stations around the world:

Area	Number of Service Stations	Petroleum Product Sales (MBPD)	Sales per Site (BPD)
Europe	9,110	1,093	120
Africa	4,167	419	100
Americas	585	76	130
Middle East	809	55	68
Asia	1,790	150	84
Total	**16,461**	**1,793**	**109**

Lubricants

The production and sales of lubricants is a highly profitable business for Total, accounting for more than one third of total Marketing & Services' earnings[367].

---

[362] TOTAL 2016 Factbook, page 126
[363] Ibid, page 105
[364] Ibid, page 121. 3,418MBPD includes petroleum products sold under Total's Refining & Chemicals segment
[365] Ibid, page 125. 1,749MBPD represents volumes entirely attributable to the Marketing & Services segment.
[366] Ibid, page 119
[367] Total 2016 Registration document, page 37

Total's lubricants operations encompass 41 blending plants around the world[368]:

- Europe: In this region Total pursues development of high-growth segments. The two main plants that supply lubricants to the region are in France and Belgium.

- Africa & Middle East: The Company relies on its production plants in Dubai, Egypt and Saudi Arabia to supply this region. In Africa Total is the leading distributor of lubricants with a 16.5% market share.

- Asia-Pacific: Total's share of the lubricant market in that region reached 3.6%. The company opened up two new production facilities in Singapore and China, respectively, which will help the company achieve its ambitions for growth in this region.

- Americas: Total is looking to grow their lubricants business in the region, primarily aviation fuels and special fluids. The company began operations in a new facility in Bayport, Texas in 2016.

Overall Company Metrics

At year-end 2016 Total had a market capitalization of approximately $124 billion, consolidated revenues of $150 billion, and adjusted earnings of $8.3 billion. Total had EBITDA of $21.8 billion and an Enterprise Value of $159.5 billion indicating an EBITDA multiple of 7. Return on Capital Employed (ROCE), using the metric provided throughout the book, was 4% while Total's provided metric was 7.5%[369]. The company's adjusted Cash Return on Capital Employed (CROCE) was 11.8%. Total had average equity of $98.5 billion[370], indicating Return on Equity (ROE) of 6.3%, while the company provided ROE metric was 8.7%[371]. Total's year-end ADR share price was $50.97 and distributed dividends per share of $2.61, indicating a dividend yield of 5% as of December 31, 2016.

Throughout the year, Total distributed $2.6 billion in dividends to its shareholders and the company issued *net shares* for proceeds of $248MM[372]. In terms of Cash Flow, Total generated $16.5 billion in cash flow from operations[373] devoting 16% to dividends during 2016 while using $18.1

---

[368] Total 2016 Form 20-F, pages 68-70
[369] TOTAL Fact Book 2016, page 7. Differences are primarily attributable to different definitions of capital employed
[370] Total Fact Book 2016, page 14
[371] Differences primarily attributable to using adjusted earnings as the numerator
[372] Total Fact Book 2016, page 22
[373] Ibid, page 22

billion for capital expenditures. Because of dividend payments and capital expenditures *exceeded* cash flow from operations; Total experienced a *negative* Free Cash Flow of $1.6 billion in 2016. With a beginning ADR share price of $44.95, $2.61 of dividends and year-end ADR share price of $50.97, Total achieved a total shareholder return (TSR) of 19.2%[374]. Total had earnings per share or EPS of $3.38 on a fully diluted basis, indicating a price earnings ratio of 15, based on a share price of $50.97. Total had total debt of $57 billion and ending equity of $102 billion, indicating a total Debt-to-Equity ratio of 56%[375]. Their net-to-debt equity ratio, which takes into account cash and cash equivalents of $25 billion, was a more modest 27%[376]. With more than $1.1 billion in interest expense and EBIT or operating income $8.3 billion in 2016, Total had an interest coverage ratio of 7.5. Total had current assets of $72.5 billion, current liabilities of $54.7 billion, translating to a current ratio of 1.326 and a *positive* working capital of $17.8 billion.

Total had about 102,168 employees at year-end 2016, indicating adjusted earnings per employee of $81M and operating cash flows per employee of $162M for the same year.

Refining, Chemicals & Marketing Metrics

Total had total throughput volumes of 1,965MBPD, with total refined products production of 1,871MBPD total refined and a global refining capacity of 2,011MBPD in 2016[377], indicating a refinery utilization of 97% in 2016. Total had total clean product sales of 1,301MBPD, which based on total throughput of 1,965MBPD translated to a clean product yield of 66%[378].

Refining & Chemicals segment had $71 billion in revenues (including inter-segment sales) in 2016[379]. Total's downstream operations had adjusted earnings of $4.6 billion and D&A of $1.0 billion[380] in 2016. Using total refined product sales of 4,183MBPD adjusted earnings per barrel were $6.72/bbl.[381] and cash per barrel was $8.18/bbl.

---

[374] Return based on ADR price and U.S. Dollar. TSR in based Euros might be different.
[375]Total Fact Book 2016, page 14
[376] Ibid, page 18
[377] Total Fact Book 2016, pages 103 & 104
[378] Based on transportation fuels of 1,301MBPD divided by throughput volumes of 1,965MBPD. See page 104 of Total Fact Book 2016.
[379] Ibid, page 11
[380] Ibid
[381] Using total refined product sales as the denominator.

Chapter VII – Asia & the Middle East

"The secret of getting ahead is getting started." – Mark Twain

Saudi Aramco (SA)

www.saudiaramco.com

"There is always room at the top" – Daniel Webster

Company Overview

Saudi Aramco is the largest oil & gas company in the world, in terms of production, earnings and as well the largest company in the world of any sector. Founded in the 1930's initially as the "California Arabian Company" by Standard Oil of California (now Chevron), SOCAL was later joined by Texaco to create the "Arab-American Company" or Aramco. Exxon and Mobil subsequently joined the Aramco joint venture. Aramco was not fully nationalized until the early 1980's and its name was subsequently changed to Saudi Aramco in 1988.

Saudi Aramco employs about 65,266 people, the majority of which are Saudi nationals[382]. Saudi Aramco is headquartered in Riyadh, Saudi Arabia and the company has offices in cities around the world, including Houston, Singapore and Beijing.

Please note that Saudi Aramco does not *externally* publish its financial statements and therefore no company financial metrics are provided. Instead, for Saudi Aramco, extensive operational overview and operational metrics are provided for the year 2015 (the latest year available at the time of publication).

Refining & Chemicals

In the downstream, Saudi Aramco is the world's sixth largest refiner[383], with a net refining capacity of 3.1MMBPD (5.4MMBPD gross). Saudi Aramco's long term goal is to become the largest refiner company in the world, with a target refining capacity of 8-10MMBPD over the next years[384]. In 2015 Saudi Aramco produced 477MBPD of gasoline, 695MBPD of diesel and 124MBPD of jet fuel in the Kingdom of Saudi Arabia. Saudi Aramco had a clean product yield of 75% in 2015 based on total transportation fuels

---

[382] Saudi Aramco 2015 Annual Review, page 98
[383] Saudi Aramco 2013 Annual Review, page 6
[384] http://www.petroleum-economist.com/articles/corporate/company-profiles/2017/saudi-aramcos-shifting-strategy

production of 1,295MBPD and total *estimated* throughput volumes of 1,723MBPD[385].

In 2015, Saudi Aramco had total net refining capacity in Saudi Arabia of 1,999MBPD and had total *estimated* throughput volumes of 1,723MBPD, indicating a refinery utilization rate of 86% in Saudi Arabia. Saudi Aramco has expanded their international downstream presence, acquiring the other 50% it did not own of Motiva Enterprises in the United States, paying close to $2.2 billion to Shell in 2017[386].

Saudi Aramco has a worldwide presence that now includes Motiva Enterprise as a fully owned subsidiary of Saudi Refining Inc. Saudi Aramco additionally has a significant presence in the chemicals business in the Kingdom of Saudi Arabia and around the world. Saudi Aramco's goals are to continue to expand further in the downstream and the chemicals businesses, as well as continue to develop unconventional gas resources to power the economic growth of the Kingdom of Saudi Arabia and the world.

The following table provides an overview of Saudi Aramco's refineries in Saudi Arabia and abroad[387]:

Refinery	Country	Total Capacity (MBPD)	Ownership	Net Capacity (MBPD)
Ras Tanura	Saudi Arabia	550	100%	550
Riyadh	Saudi Arabia	126	100%	126
Jiddah	Saudi Arabia	78	100%	78
Yanbu'	Saudi Arabia	245	100%	245
Petro Rabigh	Saudi Arabia	400	37.50%	150
SAMREF-Yanbu'	Saudi Arabia	400	50%	200
YASREF-Yanbu'	Saudi Arabia	400	62.50%	250
SASREF-Jubail	Saudi Arabia	300	50%	150
SATORP-Jubail	Saudi Arabia	400	62.50%	250
Total Saudi Arabia		**2,899**		**1,999**
Motiva Enterprises	USA	600	100%	600
S-OIL	South Korea	669	63.40%	424
Showa Shell	Japan	445	14.96%	66.65
FREP	China	280	25%	70
Total Refineries		**4,893**		**3,160**

---

[385] Saudi Aramco 2015 Annual Review, pages 96-98. Throughput volumes estimated to be 98% of total refined products production of 1,758MPD.

[386] https://www.bloomberg.com/news/articles/2017-03-07/saudi-aramco-and-shell-agree-on-terms-of-motiva-asset-breakup

[387] Saudi Aramco 2015 Annual Review, page 96

Saudi Arabian Refineries

Saudi Aramco owns and operates four domestic refineries, Jiddah, Ras Tanura, Riyadh and Yanbu refineries. A fifth refinery will be completed in 2017/2018, the Jazan refinery:

- Jiddah refinery started operations in 1967 and has a refining capacity of 78MBPD[388]. This refinery has marine and product handling facilities capable of handling 65MBPD of hydrocarbon products, as well as import and export capabilities for the associated terminals.

- Ras Tanura Refinery is Saudi Aramco's most complex refinery in the Arabian Gulf and has a crude distillation capacity of 550MBPD[389] and began operations in 1945. Additionally, the refinery has a 305MBPD NGL processing facility, a 960MBPD crude oil stabilization unit, crude oil and refined product tanks capable of handling 5.8MMBbls. Most of the products out of this refinery are sold in the domestic market[390]. This refinery is currently undergoing a capacity expansion which will add an isomerization unit, naphtha hydro treater, continuous catalytic reforming unit which will help comply with ever rising clean product specifications[391].

- Riyadh Refinery, located in the Central Region of Saudi Arabia, has a refining capacity of 126MBPD.[392]

- Yanbu Refinery became operational in 1983 and has a refining capacity of 245MBPD.

- Jazan Refinery, which is currently under construction and will commence operations in 2017/2018, will have a refining capacity of 400MBPD. This refinery will process Arabian Heavy and Medium crude oils and produce gasoline, ultra-low sulfur diesel and other products. This refinery encompasses the refinery, a terminal as well associated utilities, such as the world's largest integrated gasification combined cycle power plant, with a capacity to generate up to 4,000 MW[393].

---

[388] http://www.saudiaramco.com/en/home/our-business/worlds-leading-supplier-of-energy/refining.html
[389] Ibid
[390] Ibid
[391] http://www.ogj.com/articles/2017/01/aramco-lets-contract-for-ras-tanura-refinery.html
[392] Ibid
[393] Saudi Aramco 2015 Annual Review, page 31

Joint Ventures & Overseas Refineries

Saudi Aramco has several joint ventures in the refining and petrochemicals sectors:

- SATORP, or the Saudi Aramco Total Refining and Petrochemical Company is a joint venture with France's Total S.A. This complex is a newly constructed 400MBPD full conversion refinery with integrated petrochemical production[394]. This refinery was completed in 2013.

- SADARA, or the Sadara Chemical Company, is a joint venture with Dow Chemical Company. SADARA is an integrated chemical complex located in the Jubail Industrial City. This complex began operations in early 2016 is expected to have the ability to produce up to 3.2 million tons of chemicals per year. Sadara is the largest integrated chemicals plant ever built in a single phase. Sadara is also the first chemicals complex in the Middle East that cracks naphtha, which supports the manufacturing of diverse products. The complex includes a mixed feed cracking unit capable of processing 85MMCFD of ethane and 53MBPD of naphtha as feedstock to produce 3 million tons of performance plastics and high value chemicals per year

- YASREF, or the Yanbu Aramco Sinopec Refining Company, is a joint venture between Saudi Aramco and the Chinese company Sinopec. This refinery is Saudi Arabia's most complex refinery, with a capacity to process 400MBPD of crude oil and produce gasoline, high-quality diesel, LPGs for export, sulfur and petroleum coke for export.

- Petro Rabigh is a joint venture with Japan's Sumitomo Chemical Company. The complex is located in the town of Rabigh, on the Red Sea on Saudi Arabia's west coast. Petro Rabigh has the capacity to process up 400MBPD of crude oil and the capacity to produce a variety of transportation fuels as well as other products. Saudi Aramco supplies Petro Rabigh with 400MBPD of crude oil, 95MMCFD of ethane and about 15MBPD of butane. This feedstock is processed by a state-of-the-art plant that includes the world's largest and most sophisticated High Olefins Fluid Catalytic Cracker (HOFCC) and Ethane Cracker. The HOFCC uses high temperatures and catalysts to convert heavy oil into gasoline,

---

[394] Saudi Aramco 2013 Annual Review, page 34

naphtha and, finally, propylene, used to make everything from insulation to durable goods. The Ethane Cracker uses similar "cracking" processes to convert ethane into polymer grade ethylene, used in the manufacture of lubricants, detergents and other products. Petro Rabigh comprises 23 plants producing 18.4 million tons per year of petroleum-based products and 2.4 million tons per year of ethylene and propylene-based derivatives[395].

- SAMREF, which is the Saudi Aramco Mobil Refining Company, is a joint venture with ExxonMobil. SAMREF has the capacity to process 400MBPD of crude oil and produce different petroleum products.

- Luberef, which is the Saudi Aramco Lubricating Oil Refining, is a venture with Jadwa Industrial investment, has the capacity to produce 4 million barrels per year of base oil, and the resulting lubricant products are primarily sold in the Saudi Arabian domestic market.

- SASREF, Saudi Aramco Shell Refinery Company – located in Al Jubail is a 50/50 joint venture with Shell with a total refining capacity of 305MBPD. Saudi Aramco supplies Arabian Light Crude feedstock and natural gas by pipeline to this refinery.[396]

Sales & Marketing

Saudi Aramco Products Trading Company (ATC), a Saudi Aramco's fully owned subsidiary for trading petroleum products, traded an average of more than 1.1MMBPD of refined petroleum products and 3.7 thousand tons per day of chemical products in 2015. Over the past couple of years, ATC has expanded its market presence by reaching out directly to third parties for product sales and purchases, optimizing operations of the storage and blending facility, introducing chemical products sales and marketing, and increasing chartering activities in response to growing trade volumes.

Saudi Aramco operates a network of more than 22,000km of pipelines, as well as network of terminals and service stations.

---

[395] https://www.petrorabigh.com/en/AboutPRC/WhoWeAre/Pages/AtAGlance.aspx
[396] www.sasref.com/sa/EN/refinery.html

Gas Processing & NGL

Saudi Aramco has an extensive natural gas producing and processing assets. In 2015 Saudi Aramco produced 11.6 BCF/d of natural gas and 1.3MMBPD of NGLs. Some of Saudi Aramco's gas assets are presented here:

- Abqaiq NGL, a group of NGL facilities consisting of 8 compression trains receiving feed gas from different sources.
- Berri Gas plant, with a total capacity of 0.9 BCF/d, processes gas from the Qatif field as well as other sources. The NGL system is connected to supply the Jubail Industrial Complex
- Haradh Gas plant, located 280 km southwest of Dhahran, became operational in 2003 and has a capacity of 1.95 BCF/d of natural gas, processing natural gas primarily from the South Ghawar area.
- Hawiyah gas plant, located about 220 km south of Dhahran, became operational in 2001. This plant has a total capacity of 2.4 BCF/d.
- Hawiyah NGL recovery plant, became operational in 2008 and is designed to process 4 BCF/d and can recover up to 585MBPD of NGLs. Hawiyah is specifically designed to meet increasing demand for ethane as a petrochemical feedstock and is connected to the Yanbu and Rabigh petrochemical complexes
- Ju'aymah, which started operations in 1980, is the world's largest NGL fractionation plant, with a current capacity of 1.1MMBPD[397].
- Khursaniyah Gas plant, started operations in 2010, this complex can handle up to 2.88 BCF/d of natural gas from several fields
- Shedgum complex, started production in 1980, has the capacity to process up to 1.5 BCF/d of natural gas from several fields
- 'Uthmaniyah Gas plant, began operations in 1981, has a capacity to process associated natural gas to the tune of 1.5 BCF/d.
- Yanbu' NGL fractionator, started operations in 1982, has a current capacity to fractionate 585MBPD of NGLs.
- Karan gas program, completed in 2012, processes gas from the non-associated offshore gas field by the same name. This gas program reached its full production capacity of 1.8BCF/d a few months later.

---

[397] The Enterprise Products 1.1MMBPD NGL fractionator in Mont Belvieu ties with Ju'aymah as the world's largest.

Petrochina (PTR)

www.petrochina.com.cn

"A book holds a house of gold" – Chinese proverb

Company Overview

Petrochina is China's largest integrated oil & gas company and one of the largest companies in the world. Petrochina is engaged in exploration & production, refining, chemicals, marketing and natural gas & pipelines. China National Petroleum Corporation (CNPC) owns approximately 86.2% of Petrochina's shares[398], with the remaining shares being traded on several stock exchanges.

Petrochina was established in November 1999, in its current form as a joint stock company under the People's Republic of China (PRC), as part of a restructuring in which China National Petroleum Corporation (CNPC) transferred to Petrochina most of its assets in E&P, R&M, Chemicals and pipeline businesses[399]. Shortly after restructuring, in the year 2000, Petrochina was listed on the Hong Kong Stock Exchange as well as through ADS (American Depositary Shares) on the New York Stock Exchange.

The development of the oil and gas industry in the People's Republic of China (PRC) began in the 1950's with several major discoveries of crude oil and natural gas in Yumen, Xinjian, Qinghai and Daqing regions[400]. In 1978, the PRC became the eighth largest crude oil producer in the world.

Petrochina's NYSE ticker symbol is PTR and is headquartered in Beijing, China. At year-end 2016, Petrochina had more than 508,000 employees[401].

The effective Renminbi/Chinese Yuan to USD rate used for conversion purposes was 6.94 yuan for US$1[402].

Areas of Operation

Petrochina is a fully integrated energy company, with operations in exploration, production, transportation, refining, marketing of crude oil, refined products, natural gas, petrochemicals and other products.

---

[398] Petrochina 2016 20-F form, page 12
[399] Petrochina 2016 20-F form, page 17
[400] Petrochina IPO Prospectus, page 73
[401] Petrochina 2016 Annual Report, page 102
[402]Ibid, page 153

Refining & Chemicals

Petrochina began limited refining activities in the mid-1950s, when the company began producing gasoline and diesel at refineries in the Yumen oil region. The company now operates 29 refineries located throughout China.[403] Petrochina's Refining & Chemicals segment had revenues of approximately $84 billion in 2016[404].

Refineries

Petrochina has a crude oil refining capacity of approximately 3,446MBPD, composed of seven major petrochemical complexes and several other refineries. Petrochina's three largest refineries are Dalian, Fushun & Lanzhou. Most of Petrochina's refineries are located close to the company's crude oil production and pipeline locations, which provide refineries with secure supplies as well as facilitates the distribution of refined products. Below are the company's refining centers, their respective capacities and throughput volumes[405] as of 2016:

Refinery	Refining throughput (MBPD)	Primary distillation capacity (MBPD)	Utilization Rate %
Lanzhou Petrochemical	167	213	78%
Dalian Petrochemical	264	415	64%
Fushun Petrochemical	171	223	77%
Dushanzi Petrochemical	151	202	75%
Guangxi Petrochemical	91	202	45%
Jilin Petrochemical	184	198	93%
Sichuan Petrochemical	141	202	70%
Other refineries	1,444	1,790	81%
Total	**2,612**	**3,446**	**76%**

- Lanzhou – the first large-scale refining center in West China, this center has a 213MBPD crude distillation capacity and more than 700,000 metric tons of ethylene production capacity per year. This complex produces transportation fuels as well as petrochemicals.

- Dalian – China's largest refinery, Dalian is a joint venture between several partners and Petrochina, which holds a 28.44% interest.

- Fushun – located in northeastern China, this refinery has a crude distillation capacity of 223MBPD[406]. This refinery produces a variety of transportation fuels, as well as several petrochemical

[403] Petrochina IPO Prospectus, page 79
[404] Petrochina 2016 Annual Report, page 185
[405] Petrochina 2016 20-F form, page 31
[406] Petrochina 2016 20-F form, page 31

products, being one of the world's most important production locations for paraffin and alkyl benzene.

- Dushanzi – located in Xinjiang region, Dushanzi is one of China's largest integrated refining and petrochemical operations with a crude distillation capacity of 202MBPD[407] and an ethylene production capacity of 1.2 million metric tons per year. Construction started in 2005 and the complex began operations in 2009.

- Guangxi – located in southern China, this refinery has a crude distillation capacity of 202MBPD and was completed in 2010. The refinery produces a variety of transportation fuels as well as more than 900,000 metric tons of petrochemical products per year.

- Jilin – located in Jilin province in northeastern China, this refinery has a crude distillation capacity of 198MBPD and recently underwent an expansion in 2010.

- Sichuan- located in Sichuan province in central China, this refinery has refining operations, with a crude oil distillation capacity of 202MBPD as well as petrochemical operations capable of producing 800,000 metric tons of ethylene per year. This refinery is the newest refinery in Petrochina's portfolio and commenced operations in 2014[408].

- Yunnan Refinery, the company has plans to commence operations at this refinery in mid to late 2017 and is expected to have a crude oil distillation capacity of 261MPBD[409]. The Yunnan refinery will produce transportation fuels and other products for the domestic market as well as being able to export petroleum products to nearby South East Asian countries. This refinery will process imported crude oils, primarily piped through a Myanmar-Chinese joint pipeline project. Petrochina is currently in talks with Saudi Aramco to sell an equity portion of this refinery and deepen the ties between the world's largest crude oil producer and the world's largest crude oil demand center[410].

---

[407] Ibid

[408] https://www.icis.com/resources/news/2014/01/13/9742710/petrochina-starts-up-sichuan-refinery-and-petrochemical-complex/

[409] http://www.chinadaily.com.cn/bizchina/2017-06/02/content_29585361.htm

[410] https://www.bloomberg.com/news/articles/2017-05-17/saudi-aramco-tightens-grip-in-top-oil-market-with-china-refinery

Petrochina is one of the largest participants in China's refining market, producing about 20% of China's demand for petroleum products including gasoline, diesel, kerosene and lubricants in 2016:

Refined Product	2016 Production Volumes (MBPD)
Diesel	954
Gasoline	761
Kerosene	131
Lubricants	20
Fuel oil	39
Naphtha	227
Total	2,132

Chemicals

Petrochina is one of the major producers of ethylene in China[411]. The company produces substantial quantities of basic petrochemicals, such as ethylene, propylene & benzene as well as derivative chemicals such as resins, synthetic rubber, urea and other polymers.

Type	2016 Production Volumes (thousand tons)
Propylene	5,120
Ethylene	5,589
Benzene	1,918
Synthetic resin	9,078
Other synthetic fiber raw materials and polymer	1,410
Synthetic rubber	760
Urea	1,900
Total	25,775

Marketing

Petrochina has substantial marketing operations within China, with operations that are conducted through 36 regional sales companies. The company's marketing operations include transportation, storage, wholesale, retail and export of gasoline, diesel, kerosene, lubricants asphalt and other refined products[412]. In 2016, the company's marketing operations achieved a record 3.4MMBPD of refined petroleum sales[413]:

Product	Barrels MBPD
Diesel Sales	1,639
Gasoline Sales	1,428
Kerosene Sales	357
Lubricants Sales	20
Total Sales	3,443

---

[411] http://www.cnpc.com.cn/en/refiningchemicals/common_index.shtml
[412] Petrochina 2016 20-F form, page 34
[413] Ibid

Marketing operations had revenues of $318 billion and operating income of $1.6 billion. The company has both wholesale and retail operations, and their retail presence encompasses approximately 20,895 service stations of which more than 96% are company operated; the remaining are franchised businesses. Capital expenditures in 2016 were $1.6 billion, primarily geared towards opening new company owned & operated service stations.

The company's retail marketing group achieved an average of 77BPD or 10.5 tons per year of sales per service station.

Overall Company Metrics

Please note that Petrochina's financial statements are published in Chinese Renminbi. The exchange rate used to translate the financial statements to U.S. dollars was the exchange referenced in chapter one of 6.94RMB or Chinese Yuan for one US dollar.

At the end of 2016, Petrochina had a market cap of $135 billion, Enterprise Value of $222 billion, consolidated revenues of $233 billion[414], adjusted earnings of $1.9 billion and EBIT or *profit from operations* of $8.7 billion. With an EBITDA of $40 billion, Petrochina's EV over EBITDA multiple was 5.5. In 2016, Return on Capital Employed for Petrochina was 2.1%, while Cash return on capital employed for the same period was 13.2%, and Return on Equity 2%. For the same period, Petrochina distributed dividends of $1.2 billion, which based on a year-end share price of $73.70 and yearly dividends in 2016 of $0.48/share, equate to an effective dividend yield of roughly 0.65%. The company's effective income tax rate in 2016 was 35%. Earnings per share were $0.62 indicating that the company's shares were trading at a 119 price-earnings ratio. Petrochina paid a total of $2.8 billion in interest in 2016 while its EBIT was $8.7 billion, translating to an interest coverage ratio of 3.1.

In 2016, Petrochina generated $38.2 billion in cash flow from operations, devoting approximately 3% of those operating cash flows to pay dividends in the same year. Petrochina's capital expenditures in 2016 were $26.1 billion, indicating that free cash flow was $12.1 billion. For the same year, Petrochina had a 12.78% total shareholder return. Petrochina, at year-end 2016, had debt of approximately $74 billion and total equity of $198 billion[415], indicating a debt-to-equity ratio of 37%. Current assets were $55

---

[414] Petrochina 2016 20-F form, page F-4
[415] Ibid, page F-5

billion, while current liabilities were $72 billion translating to a *negative* working capital of $17 billion and a current ratio of 0.764.

Based on 508M employees, adjusted earnings of $1.9 billion, cash flow from operations of $38 billion, adjusted earnings per employee were $4M, while operating cash flow per employee was $75M.

Downstream Metrics

The company has a global refining capacity of 3.4MMBPD, with the majority of the capacity located in China. The largest refineries for the company, based on primary distillation capacity, were the Dalian, Fushun and Lanzhou petrochemical complexes. Total crude oil processed was 2.6MMBPD[416]. The company achieved a refinery utilization of 80%[417] in 2016 and had a clean product yield of 68%[418] for the same period.

In 2016, Petrochina's Refining, Chemicals & Marketing segment had EBIT of $7.2 billion[419]. Earnings per barrel were $9.60, while cash per barrel was $16.95.

---

[416] Petrochina 2016 20-F form, page 30
[417] Petrochina 2016 20-F form, page 31
[418] Ibid
[419]http://www.petrochina.com.cn/ptr/xwxx/201404/1124915c84d64162955f1edcfb12c6cf.shtml

Chapter VIII – Latin America

"If you want to conquer fear, don't sit home and think about it. Go out and get busy." – Dale Carnegie

Petrobras (PBR, PBR-A)

www.petrobras.com

"A fall into a ditch makes you wiser" – Chinese proverb

Company Overview

Petrobras was founded in October 1953 as Petróleo Brasileiro[420]; the company today is one of the largest publicly-traded national oil companies in the world. Petrobras is a fully integrated energy company, engaged in exploration & production, refining & marketing/retail, gas & power generation and biofuels.

Petrobras was the Brazilian federal government's exclusive agent and fully national company between 1953 and 1997, but after several reforms, including an amendment to the Brazilian constitution, a concession system was established in Brazil. This concession system allowed foreign company participation in the Brazilian oil & gas sector. Petrobras is one of the recent success stories in the deep and ultra-deep water exploration and production areas[421]. Petrobras crude oil production has grown by a factor of 10 between 1980 and 2013, growing from 181MPBD to 2,790MBPD[422].

Petrobras is listed on the Brazilian Stock exchange (BOVESPA) as well as on the NYSE under the ticker symbol PBR (common stock) and PBR-A (preferred stock). The Brazilian government owns 28.67% of all Petrobras capital stock, while domestic and foreign investors own about 51.68% with the remaining shares owned by different Brazilian institutions[423]. Shares of Petrobras (both common and preferred) have been trading on the BOVESPA stock exchange since 1968 and on the NYSE since the year 2000[424].

Petrobras is headquartered in Rio de Janeiro, Brazil and at year-end 2016 had more 68,000 employees[425].

---

[420] Petrobras 2016 Form 20-F, page 39
[421] http://www.ogj.com/articles/2014/10/petrobras-makes-deepwater-gas-condensate-discovery-in-espirito-santo-basin.html
[422] 2016 Report to the Petrobras Administration, page 14
[423] 2016 Report to the Petrobras Administration, page 10
[424] Petrobras 2016 Form 20-F, page 146
[425] 2016 Report to the Petrobras Administration, page 74

Areas of Operation

Petrobras is one of the world's largest integrated oil & gas companies, with primary operations in Brazil, being the main participating in the hydrocarbon sector in that country. The company operates most of Brazil's oil & gas production as well as operates substantially almost all of Brazil's refining capacity[426].

Petrobras has operations in 18 countries, with E&P operations in Brazil, US, South America and Africa, and with downstream operations in Brazil, South America, US, Africa and Japan.

The company is organized into five business segments[427]:

- Exploration and Production: this segment covers the activities of exploration, development and production of crude oil, LNG and natural gas in Brazil and abroad, for the primary purpose of supplying Petrobras' domestic refineries and selling surplus crude oil and oil products produced in the natural gas processing plants to the domestic and foreign markets. The E&P segment also operates through partnerships with other companies.
- Refining, Marketing and Transportation: this segment covers the activities of refining, logistics, transport and trading of crude oil and oil products in Brazil and abroad, exports of ethanol, extraction and processing of shale, as well as holding interests in petrochemical companies in Brazil.
- Gas and Power: this segment covers the activities of transportation and trading of natural gas produced in Brazil and abroad, imported natural gas, transportation and trading of LNG, generation and trading of electricity, as well as holding interests in transporters and distributors of natural gas and in thermoelectric power plants in Brazil, in addition to being responsible for the fertilizer business.
- Distribution: this segment covers the activities of Petrobras Distribuidora S.A, which sells oil products, ethanol and vehicle natural gas in Brazil.
- Biofuel: this business segment covers the activities of production of biodiesel and its co-products, as well as ethanol-related activities such as equity investments, production and trading of ethanol,

---

[426] Petrobras 2016 Form 20-F, page 40
[427] Ibid, page 41

sugar and the surplus electric power generated from sugarcane bagasse.

Refining, Marketing & Transportation

Petrobras currently operates 13 refineries in Brazil and one refinery in the United States[428]:

Name	Location	Crude Distillation Capacity (MBPD)	2016 Average Throughput (MBPD)	2015 Average Throughput (MBPD)
LUBNOR	Fortaleza, Brazil	8	9	8
RECAP (Capuava)	Capuava, Brazil	53	54	40
REDUC (Duque de Caxias)	Duque de Caxias, Brazil	239	194	235
REFAP (Alberto Pasqualini)	Canoas, Brazil	201	148	174
REGAP (Gabriel Passos)	Betim, Brazil	157	150	152
REMAN (Isaac Sabbá)	Manaus, Brazil	46	34	38
REPAR (Presidente Getúlio Vargas)	Araucaria, Brazil	208	167	197
REPLAN (Paulínia)	Paulinia, Brazil	415	331	391
REVAP (Henrique Lage)	Sao Jose dos Campos, Brazil	252	217	249
RLAM (Landulpho Alves)	Mataripe, Brazil	315	218	248
RPBC (Presidente Bernardes)	Cubatao, Brazil	170	142	157
RPCC (Potiguar Clara Camarão)	Guamare, Brazil	38	33	34
RNEST (Abreu e Lima)	Ipojuca, Brazil	74	75	53
Pasadena Refining System	Pasadena, TX, USA	100	104.2	99.5
Total		**2,276**	**1,876**	**2,076**

- Paulínia or Replan is Petrobras' largest refinery, with a crude distillation capacity of 415MBPD and the production of refined products from this refinery corresponds to 20% of all refining capacity in Brazil[429]. This refinery processes primarily domestic crude oil most of which coming from the Campos Basin. This refinery was built in 1972 and processes 126MBPD of crude oil. This refinery supplies several nearby Brazilian states with a variety of products, including diesel, gasoline, LPG, fuel oils, jet fuel asphalt, propene and other products.

---

[428] Petrobras 2016 Form 20-F, page 61
[429] http://www.petrobras.com.br/en/our-activities/main-operations/refineries/paulinia-replan.htm

- Landulpho Alves or RLAM was the first Brazilian oil refinery, built in 1950 and is currently the second largest refinery with a crude distillation capacity of 315MPD. This refinery enabled the development of the first planned petrochemical complex in Brazil and the largest industrial complex in the Southern Hemisphere, the Camaçari Petrochemical Complex[430]. This refinery produces a variety of transportation fuels, as well as lubricants, petrochemical feedstocks used in a wide variety of applications, from chewing gum to detergents.

- Henrique Lage or REVAP, is Brazil's third largest refinery, with a crude distillation capacity of 252MBPD, was built in 1980. This refinery can produce a wide variety of products, from transportation fuels to asphalts, coke, sulfur, LPGs and petrochemical feedstocks[431].

- Duque de Caxias or REDUC is a large refinery with a distillation capacity of 239MBPD, was built in 1961 which accounts for 80% of the lubricant production in the country as well as having the largest natural gas processing in the country[432]. This refinery supplies nearby Brazilian states with a wide variety of products, being the refinery in Petrobras' portfolio with the most different types of finished products and feedstocks.

- Presidente Getúlio Vargas or REPAR is the fifth largest refinery in Brazil with a crude distillation capacity of 208MBPD and ships 85% of its production to nearby states. The refinery was built in 1977 and produces transportation fuels as well as marine oils[433].

- Alberto Pasqualini or REFAP has a capacity of 201MBPD and was built in 1968. This refinery is equipped with several advanced units, including hydrodesulphurization units, power co-generation unit, solvent unit, propene unit and is designed to maximize diesel production, including low-sulfur diesel. .

- Presidente Bernardes or RPBC is a refinery with a high conversation capacity, producing dozens of high market value

[430] http://www.petrobras.com.br/en/our-activities/main-operations/refineries/landulpho-alves-rlam.htm

[431] http://www.petrobras.com.br/en/our-activities/main-operations/refineries/henrique-lage-revap.htm

[432] http://www.petrobras.com.br/en/our-activities/main-operations/refineries/duque-de-caxias-reduc.htm

[433] http://www.petrobras.com.br/en/our-activities/main-operations/refineries/presidente-vargas-repar.htm

international standard petroleum products, has a crude distillation capacity of 170MBPD and was built in 1951.

- Gabriel Passos or REGAP was inaugurated in 1968 and currently has a distillation capacity of 157MBPD. The refinery has undergone a series of upgrades, including a new diesel fuel unit, naphtha hydro desulfurization, coker, LPG spheres and many other upgrades. This refinery produces a variety of products, including gasoline, diesel, jet fuel, bunker fuel, green petroleum coke and others.

- Pasadena Refining System is a refinery in the city of Pasadena, Texas and has a crude oil distillation capacity of 100MBPD. Petrobras acquired the refinery in 2012, but the refinery was originally built in 1920 as a lubricants plant[434]. The refinery has been modernized over the years and has several units, including catalytic cracking, catalytic reforming, desulfurization units as well as vacuum distillation units.

- Abreu e Lima or RNEST currently has a crude oil capacity of 74MBPD, but has a project to expand that capacity to be able to process up to 230MBPD[435]. The expansion project is already in progress with Train 1 complete and Train 2 of this expansion still underway. RNEST will have the highest crude oil conversion in diesel (70%) and will be the most advanced refinery in the company's portfolio. Moreover, this refinery will be able to produce refined product products that meet the strictest international standards, including 10 parts per million or *PPM* for ultra-low sulfur diesel.

- Capuava or RECAP was originally built in 1947 and has a current distillation capacity of 53MBPD. This refinery processes approximately 90% of Brazilian crude oils and produces gasoline, low sulfur diesel, propylene, LPGs and special solvents.

- Isaac Sabbá or REMAN was built in 1957 and is located in the Brazilian state of Amazonas. This refinery has a crude oil distillation capacity of 46MBPD and can produce a variety of products, including aviation kerosene, jet fuel, gasoline, fuel oils and gasoline.

---

[434] http://www.petrobras.com/en/countries/u-s-a/operations/
[435] http://www.petrobras.com.br/en/our-activities/main-operations/refineries/abreu-e-lima-refinery.htm

- Potiguar Clara Camarão or RPCC is a small refinery with a distillation capacity of 38MBPD that produces diesel, naphtha, jet fuel and gasoline located in the Rio Grande do Norte state in Brazil.

- Lubricantes e Derivados do Nordeste or LUBNOR is small refinery with a crude distillation capacity of 8MBPD and is the only refinery in Brazil capable of producing naphthenic lubricants that are used in wide variety of applications[436]. All of the crude oil that this refinery uses is heavy oil and produces asphalts and lubricating oils.

Refined Petroleum Product Sales

Petrobras' imports and exports of oil products depend on the company's refinery output and Brazilian demand levels. Much of the crude oil the company produces in Brazil is intermediate, therefore the company must import some light crude to balance the crude slates that the company's refineries use, and export mainly intermediate crude oil from Petrobras's equity E&P production in Brazil. Petrobras also imports oil products to balance any shortfall between production from Brazilian refineries and the market demand for each product. Petrobras' main refined petroleum product export is fuel oil, which the local market is oversupplied of.

The Brazilian domestic market grew rapidly from 2010 to 2014, in parallel with Brazil's economic expansion and the increase of average income, increasing by an average of 5.6%. In the last two years, as a result of the Brazilian economic slowdown, the domestic growth rate in consumption of oil products, particularly diesel, decreased as compared to the higher rates of growth experienced in prior years[437].

---

[436] http://www.petrobras.com.br/en/our-activities/main-operations/refineries/lubrificantes-e-derivados-do-nordeste-lubnor.htm
[437] Petrobras 2016 Form 20-F, page 58

The following table from Petrobras's Form 20-F provides an overview of the different petroleum products being sold and exported from Brazil:

Domestic Sales Volumes and Exports from Brazil, MBPD			
Year	2016	2015	2014
Diesel	780	923	1,001
Gasoline	545	553	620
Fuel oil	67	104	119
Naphtha	151	133	163
LPG	234	232	235
Jet fuel	101	110	110
Others	186	179	210
Total oil products	2,064	2,234	2,458
Ethanol, nitrogen fertilizers, renewables and other products	112	123	99
Natural Gas	333	432	446
Total domestic market	2,509	2,789	3,003
Exports	554	510	393
Total domestic market and exports	**3,063**	**3,299**	**3,396**
Exports			
Crude Oil	387	360	232
Fuel oil	119	125	128
Gasoline	10	3	0
Others	26	21	30
Total Exports	**542**	**509**	**390**
Imports			
Crude oil	136	277	392
Diesel	13	78	185
LPG	72	67	28
Gasoline	32	28	41
Naphtha	105	51	88
Others	16	32	29
Total imports	**374**	**533**	**805**

Distribution Segment

Petrobras is Brazil's leading oil products distributor, operating through a company-owned own retail network, wholesale channels, and by supplying other fuel wholesalers and retailers. The Distribution segment sells petroleum products that are primarily produced by the company's Refining segment, supplying a variety of petroleum products in Brazil's domestic market, including, gasoline, diesel, jet fuel, aviation kerosene, as well as LPG, natural gas, ethanol and biodiesel.

The company supplies and operates *Petrobras Distribuidora*, which accounts for 31.1% of the total Brazilian retail and wholesale distribution market. Petrobras Distribuidora distributes oil products, ethanol, biodiesel and natural gas to retail, commercial and industrial customers. In 2016, Petrobras Distribuidora sold the equivalent of 785MBPD of petroleum products and other fuels to wholesale and retail customers, of which the largest portion (40.3%) was diesel.

At December 31, 2016, the company's branded service station network was Brazil's leading retail marketer, with 8,176 service stations, or 20% of the stations in Brazil. The company owned and franchised stations that represented 25.4% of Brazil's retail sales of diesel, gasoline, ethanol, vehicular natural gas and lubricants in 2016[438].

As of December 31, 2016, the company's onshore and offshore, crude oil and oil products pipelines extended over 4,796 miles. Petrobras operates 27 marine storage terminals and 20 other tank farms with nominal aggregate storage capacity of 64.6MMBBL. The company's marine terminals handle an average of 8,981 tankers and oil barges annually. Below is the full listing of terminals and pipelines owned by Petrobras at the end of 2016[439]:

Angra dos Reis Terminal - ORBIG Pipeline	Cubatão Terminal - Pipelines between Santos and Cubatão	Macapá Terminal	Rio Grande Terminal
Aracaju Terminal	Guamaré Terminal	Maceió Terminal - OPMAC Pipeline	Santos Terminal - Pipelines between Santos and Cubatão
Barra do Riacho Waterway Terminal	Guanabara Bay Regasification Terminal (LNG)	Madre de Deus Terminal - Pipelines between Madre de Deus and RLAM	São Caetano do Sul Terminal - OBATI Pipeline
Barueri Terminal - OBATI Pipeline	Guaramirim Terminal - OPASC Pipeline	Manaus Terminal	São Francisco do Sul Terminal - OSPAR Pipeline
Belém Terminal	Guararema Terminal - OSRIO Pipeline	Mucuripe Terminal	São Luis Terminal
Biguaçu Terminal - OPASC Pipeline	Guarulhos Terminal - OSVAT 16 Pipeline	Natal Terminal	São Sebastião Terminal - OSPLAN I Pipeline
Brasília Terminal - OSBRA Pipeline	Ilha d'Água Terminal - Pipelines between Ilha d'Água and Reduc	Niterói Terminal - ORNIT Pipeline	Senador Canedo Terminal - OSBRA Pipeline
Cabedelo Terminal	Ilha Redonda Terminal - LPG-Reduc Pipeline	Norte Capixaba Terminal	Suape Terminal
Cabiúnas Terminal - OSDUC Pipeline	Itabuna Terminal - ORSUB Pipeline	Osório Terminal - OSCAN Oil Pipeline	Uberaba Terminal - OSBRA Pipeline
Campos Elíseos Terminal - ORBEL I Pipeline	Itajaí Terminal - OPASC Pipeline	Paranaguá Terminal	Uberlândia Terminal - OSBRA Pipeline
Candeias Terminal - BECAN 6 and 8 Pipeline	Japeri Terminal - OSVOL Pipeline	Pecém Regasification Terminal (LNG)	Vitória Terminal
Coari Terminal - ORSOL I and II Pipelines	Jequie Terminal - ORSUB Pipeline	Ribeirão Preto Terminal - OSBRA Pipeline	Volta Redonda Terminal - OSVOL Pipeline

[438] Petrobras 2016 Form 20-F, page 146
[439] http://www.petrobras.com.br/en/our-activities/main-operations/terminals-and-pipelines/

Overall Company Metrics

Petrobras ended 2016 with a market cap of approximately $67 billion, an enterprise value of $164 billion with a December 31st share price of $8.81 per preferred stock in PBRA and $10.11 for common stock in PBR. For the same year Petrobras had consolidated revenues of $81 billion, adjusted EBITDA of $24.5 billion, a GAAP *loss* of $4.8 billion and adjusted earnings of $1.4 billion[440]. Using $164 billion for the enterprise value and adjusted EBITDA of $24.5 billion, indicates that the market valued Petrobras at an EV/EBITDA multiple of 6.7 at the end of 2016.

Adjusted Return on Capital Employed for the same period was 3.54%, CROCE was 10.1% and Return on Equity (ROE) was *negative* 6%, primarily due to asset impairments of $6 billion in 2016. In 2016 Petrobras did not issue dividends, due to the ongoing challenges in market conditions in Brazil and in particular for the company's upstream operations with the low price of oil. The company's effective income tax rate was 19% in 2016, one of the lowest rates in the world for an integrated oil & gas company. Petrobras had interest expense of $7 billion while adjusted EBIT was $10.5 billion, indicating an interest coverage ratio of 1.5. With earnings per share a *negative* $0.37 and a year-end common share price of $8.81, the price earnings ratio was a *negative* 27.

Petrobras generated $26.1 billion in cash flow from operations, using about $14.1 billion for capital expenditures, indicating free cash flow of $12 billion in 2016. In 2016, Petrobras' common and preferred shares experienced a significant increase in price, with common shares having a TSR of 135% while preferred shares had a TSR of 159% in the NYSE[441]. Petrobras had total debt of $118 billion and total ending equity of $77.6 billion, indicating a debt-to-equity ratio of 153%[442]. Petrobras, at year-end 2016, had total cash and cash equivalents of $21.2 billion, and a current ratio of 1.8 and a working capital of $19.9 billion.

In terms of employees, the company had 68,829 employees at year-end 2016, indicating adjusted earnings of $20M per employee and cash flow from operations per employee of $379M.

---

[440] Adjusted for asset impairments of $6.2 billion in 2016
[441] Using share prices on the NYSE for PBR and PBR-A.
[442] Petrobras 2016 Form 20-F, page F-5, Statement of Financial Position

Downstream Metrics

Petrobras owns substantially all of the refining capacity in Brazil, having a domestic crude distillation capacity of 2,176MBPD, plus the company's US refinery in Pasadena, which has a capacity of 100MBPD. Total throughput volumes were 1,923MBPD, of which 1,772MBPD were crude oil processed while the remaining 47MBPD were NGL volumes, indicating a refinery utilization rate of 84%[443]. The company, in 2016, produced a total of 1,887MBPD of refined products, of which 1,319MBPD were gasoline, jet fuel and diesel volumes, indicating a clean product yield of 69%. Domestic petroleum product sales were 2,064MBPD in 2016, while renewable product sales were 112MBPD in the same period. The company had 8,564 service stations that sold the equivalent of 785MBPD, indicating an average volume sold per station of 92BPD. The distribution segment had earnings of $67MM, indicating earnings per barrel of $0.23 based on yearly product sales of 287 million barrels.

Petrobras downstream businesses, including distribution, reported earnings of $5.7 billion in 2016, adjusted earnings of $6.9 billion[444] and estimated D&A of $3.5 billion. With total yearly refined products production volumes of 689 million barrels, these metrics translated to $8.34 in earnings per barrel, $10.09 in adjusted earnings per barrel and $15.15 in cash per barrel.

Recent prominent downstream projects in refining are the Comperj and RNEST refineries. RNEST began operations at year-end 2014 and is designed to process 115MBPD while Comperj is still being assessed for its business model and possible partnerships.

---

[443] Petrobras 2016 Form 20-F, pages 56-57
[444] Downstream earnings adjusted for a $1.2 billion impairment associated with the RNEST and COMPERJ refining assets.

Comparative Tables

"Problems are not stop signs, they are guidelines." – Robert H. Schuller

CT1 – Refining Capacity and Throughput volumes

Company	Global Refining Capacity (BPD)	Total Throughput Volumes (BPD)	Refinery Utilization %
ExxonMobil	4,971,000	4,269,000	86%
Petrochina	3,445,479	2,611,781	76%
Rosneft	2,370,144	2,033,503	86%
Petrobras	2,276,000	1,923,000	84%
Total	2,011,000	1,965,000	98%
Saudi Aramco	1,999,000	1,723,039	86%
Lukoil	1,649,058	1,295,000	79%
Tesoro	895,000	825,000	92%
Delek US Refining	155,000	148,039	96%
Alon USA	147,000	139,243	95%
Total Companies Covered in this book	**19,918,682**	**16,932,605**	**85%**
% of total world's refining capacity	**20%**		

CT2 – Assessed Available Capacity as of January 2016

In MMBPD. Source: OPEC World Oil Outlook 2016

Unit	U.S. & Canada	Latin America	Africa	Europe	Russia & Caspian	Middle East	China	Other AP	World
Distillation									
Crude oil (Atmospheric)	20.0	8.0	4.2	17.1	6.6	9.5	13.2	19.0	97.5
Vacuum	9.1	3.6	1.0	6.7	2.8	2.7	5.2	5.8	37.0
Upgrading									
Coking	3.0	0.8	0.1	0.7	0.3	0.3	1.9	0.9	8.0
Visbreaking	0.2	0.4	0.2	1.6	0.5	0.6	0.2	0.5	4.1
Solvent deasphalting	0.4	0.1	0.0	0.1	0.0	0.2	0.1	0.1	1.0
Catalytic cracking	6.1	1.6	0.3	2.4	0.6	0.8	3.0	2.8	17.6
Hydrocracking	2.1	0.2	0.2	2.0	0.3	0.9	1.6	1.5	8.9
Gasoline quality									
Reforming	4.2	0.7	0.5	2.6	0.9	1.0	1.1	2.7	13.8
Isomerization	0.8	0.1	0.1	0.6	0.3	0.4	0.2	0.2	2.6
Alkylation	1.2	0.2	0.0	0.2	0.0	0.1	0.0	0.3	2.1
MTBE/ETBE	0.1	0.0	0.0	0.1	0.0	0.0	0.1	0.1	0.3
Desulphurization									
Naphtha	5.1	0.9	0.6	3.3	1.0	1.4	1.1	3.1	16.5
Gasoline	2.5	0.5	0.1	0.6	0.1	0.2	0.7	1.0	5.7
Middle distillates	6.0	2.1	0.8	5.8	1.6	2.4	3.0	6.0	27.7
Vacuum gasoil/Residual	2.9	0.4	0.0	1.8	0.2	0.5	0.4	2.8	9.0
Sulphur (short tonnes/day)	40,524	7,738	3,634	19,103	5,670	14,892	8,594	30,898	131,052
Hydrogen (MMCFD)	6,064	1,413	407	4,580	829	2,688	2,656	5,545	24,181

CT3 – Adjusted Earnings, Cash per Barrel and Clean Product Yield

Company	Downstream Adjusted Earnings per Barrel	Downstream Cash per Barrel	Clean Product Yield Percentage
Petrobras	$10.09	$15.15	69%
Petrochina	$9.60	$16.95	79%
Total	$6.72	$8.18	66%
Lukoil	$3.13	$6.23	64%
ExxonMobil	$2.10	$2.86	95%
Tesoro	$0.92	$2.75	92%
Combined Delek US	$(0.23)	$1.66	89%
Saudi Aramco	$-	$-	75%

- Adjusted earnings per barrel are calculated as: GAAP earnings, *plus* non-cash special items to arrive at total adjusted earnings.
- Cash is calculated as adjusted earnings *plus* depreciation & amortization expenses.
- Adjusted earnings or total cash generated are then divided by total throughput volumes or total refined product sales to arrive at adjusted earnings per barrel or cash per barrel.
- Clean Product Yield is calculated as the sum of all gasoline, diesel and jet fuel volumes *divided* by total throughput volumes.

CT4 – Alphabetical List of Companies

Company	Headquarter Country	Number of Employees
Alon USA	USA/Texas	2,830
Combined Delek US	USA/Tennessee	4,156
Delek US Refining	USA/Tennessee	1,326
ExxonMobil	USA/Texas	71,100
Lukoil	Russia	65,500
Petrobras	Brazil	68,829
Petrochina	China	508,000
Rosneft	Russia	295,800
Saudi Aramco	Saudi Arabia	65,266
Tesoro	USA/Texas	6,300
Total Companies Covered in this Book (Ex. Combined Delek)		**1,084,951**

CT5 - Number of Vehicles per Country

Source: http://apps.who.int/gho/data/node.main.A995

Rank	Country	Number of Vehicles	Rank	Country	Number of Vehicles
1	USA	265,043,362	51	Libya	3,553,497
2	China	250,138,212	52	Bulgaria	3,502,771
3	India	159,490,578	53	Morocco	3,286,421
4	Indonesia	104,211,132	54	New Zealand	3,250,066
5	Japan	91,377,312	55	Dominican Republic	3,215,773
6	Brazil	81,600,729	56	Denmark	2,911,147
7	Germany	52,391,000	57	Israel	2,850,513
8	Italy	51,269,218	58	UAE	2,674,894
9	Russian Federation	50,616,163	59	Slovakia	2,622,939
10	France	42,792,103	60	Guatemala	2,562,925
11	Viet Nam	40,790,841	61	Ireland	2,482,557
12	UK	35,582,650	62	Cambodia	2,457,569
13	Mexico	35,005,913	63	Serbia	2,130,035
14	Spain	32,616,105	64	Bangladesh	2,088,566
15	Thailand	32,476,977	65	Kenya	2,011,972
16	Iran	26,866,457	66	Uruguay	1,991,836
17	Poland	24,875,717	67	Lithuania	1,984,496
18	Malaysia	23,819,256	68	Croatia	1,869,370
19	Republic of Korea	23,150,619	69	Kuwait	1,841,416
20	Argentina	23,120,241	70	Costa Rica	1,759,341
21	Canada	22,366,270	71	Tunisia	1,735,339
22	Turkey	17,939,447	72	Ecuador	1,721,206
23	Australia	17,180,596	73	Lebanon	1,680,011
24	South Africa	9,909,923	74	Burkina Faso	1,545,903
25	Colombia	9,734,565	75	Ghana	1,532,080
26	Netherlands	9,612,273	76	Tanzania	1,509,786
27	Pakistan	9,080,437	77	Laos	1,439,481
28	Greece	8,035,423	78	Slovenia	1,395,704
29	Philippines	7,690,038	79	Honduras	1,378,050
30	Czech Republic	7,689,730	80	Jordan	1,263,754
31	Algeria	7,308,539	81	Uganda	1,228,425
32	Egypt	7,037,954	82	Paraguay	1,227,469
33	Belgium	6,993,767	83	Bolivia	1,206,743
34	Saudi Arabia	6,599,216	84	Yemen	1,201,890
35	Austria	6,384,971	85	Nepal	1,178,911
36	Portugal	6,056,856	86	Azerbaijan	1,135,936
37	Romania	5,985,085	87	Liberia	1,085,075
38	Finland	5,862,216	88	Oman	1,082,996
39	Nigeria	5,791,446	89	Panama	1,004,669
40	Sweden	5,755,952	90	Singapore	974,170
41	Switzerland	5,693,642	91	Kyrgyzstan	958,187
42	Sri Lanka	5,203,678	92	Georgia	951,649
43	Iraq	4,515,041	93	Zimbabwe	927,129
44	Myanmar	4,310,112	94	Bosnia/Herzegovina	881,200
45	Peru	4,264,114	95	Turkmenistan	847,874
46	Chile	4,263,084	96	Latvia	826,469
47	Kazakhstan	3,926,487	97	El Salvador	817,972
48	Belarus	3,900,442	98	Estonia	763,975
49	Hungary	3,690,599	99	Moldovia	706,785
50	Norway	3,671,885	100	Mongolia	675,064

Rank	Country	Number of Vehicles
101	Afghanistan	655,357
102	Qatar	647,878
103	Cyprus	644,068
104	Cuba	628,155
105	Chad	622,120
106	CÃ´te d'Ivoire	594,071
107	Angola	581,530
108	Nicaragua	566,731
109	Bahrain	545,155
110	Mozambique	542,336
111	Zambia	534,532
112	Botswana	520,793
113	Jamaica	518,239
114	Ethiopia	478,244
115	Albania	445,956
116	Mauritius	443,495
117	Malawi	437,416
118	Luxembourg	431,245
119	Mauritania	416,190
120	Tajikistan	411,548
121	Macedonia	403,339
122	Senegal	401,910
123	Congo	350,000
124	Malta	322,960
125	Sudan	320,974
126	Niger	315,600
127	Mali	289,828
128	Namibia	280,583
129	Iceland	245,949
130	Madagascar	219,576
131	Suriname	207,161
132	Montenegro	201,229
133	Gabon	195,000
134	Swaziland	180,103
135	Bahamas	144,388
136	Lesotho	122,997
137	Barbados	112,118

Rank	Country	Number of Vehicles
138	Congo	110,438
139	Rwanda	107,411
140	Papua New Guinea	94,297
141	Fiji	86,535
142	Andorra	76,394
143	Eritrea	70,319
144	Sierra Leone	68,802
145	Bhutan	68,173
146	Timor-Leste	63,553
147	Guinea-Bissau	62,239
148	Maldives	61,412
149	Somalia	59,457
150	Togo	58,111
151	Cabo Verde	56,690
152	San Marino	54,606
153	Gambia	54,471
154	Solomon Islands	45,000
155	Monaco	41,055
156	CAR	37,475
157	Benin	34,914
158	Guinea	33,943
159	Antigua and Barbuda	29,989
160	Saint Vincent and the Grenadines	28,368
161	Dominica	24,620
162	Seychelles	18,606
163	Samoa	17,449
164	Guyana	15,694
165	Vanuatu	14,000
166	Cook Islands	12,453
167	Micronesia (Federated States of)	8,337
168	Tonga	8,154
169	Palau	7,102
170	Kiribati	3,452
171	Marshall Islands	2,116
172	Saint Lucia	1,569

CT6 – Vehicles per 1,000 People

Source: http://apps.who.int/gho/data/node.main.A995

Rank	Country	Vehicles per 1,000 People
1	San Marino	1736
2	Monaco	1085
3	Finland	1080
4	Andorra	964
5	Italy	841
6	USA	828
7	Luxembourg	813
8	Malaysia	802
9	Malta	753
10	Austria	752
11	Iceland	746
12	Australia	736
13	Norway	728
14	Greece	722
15	New Zealand	721
16	Japan	719
17	Czech Republic	719
18	Switzerland	705
19	Spain	695
20	Slovenia	674
21	France	666
22	Lithuania	658
23	Poland	651
24	Canada	636
25	Germany	633
26	Belgium	630
27	Cook Islands	604
28	Sweden	601
29	Estonia	593
30	Uruguay	585
31	Netherlands	574
32	Libya	573
33	Portugal	571
34	Cyprus	564
35	UK	564
36	Argentina	558
37	Kuwait	547
38	Ireland	537
39	Denmark	518
40	Bulgaria	485
41	Thailand	485
42	Slovakia	481
43	Republic of Korea	470
44	Viet Nam	445
45	Croatia	436
46	Indonesia	417
47	Belarus	417
48	Bahrain	409
49	Brazil	407
50	Latvia	403

Rank	Country	Vehicles per 1,000 People
51	Barbados	394
52	Suriname	384
53	Bahamas	383
54	Hungary	371
55	Israel	369
56	Costa Rica	361
57	Mauritius	356
58	Russian Federation	354
59	Lebanon	348
60	Iran	347
61	Dominica	342
62	Palau	340
63	Antigua and Barbuda	333
64	Montenegro	324
65	Dominican Republic	309
66	Qatar	299
67	Oman	298
68	UAE	286
69	Mexico	286
70	Romania	276
71	Panama	260
72	SV /Grenadines	259
73	Botswana	258
74	Liberia	253
75	Sri Lanka	245
76	Chile	242
77	Turkey	239
78	Kazakhstan	239
79	Mongolia	238
80	Bosnia/ Herzegovina	230
81	Saudi Arabia	229
82	Serbia	224
83	Georgia	219
84	Laos	213
85	Moldova	203
86	Colombia	201
87	Seychelles	200
88	Macedonia	191
89	South Africa	188
90	Algeria	186
91	Jamaica	186
92	China	181
93	Paraguay	180
94	Singapore	180
95	Maldives	178
96	Jordan	174
97	Kyrgyzstan	173
98	Honduras	170
99	Guatemala	166
100	Cambodia	162

Rank	Country	Vehicles per 1,000 People
101	Turkmenistan	162
102	Tunisia	158
103	Swaziland	144
104	Albania	141
105	Peru	140
106	Iraq	134
107	El Salvador	129
108	India	127
109	Namibia	122
110	Azerbaijan	121
111	Gabon	117
112	Cabo Verde	114
113	Bolivia	113
114	Ecuador	109
115	Mauritania	107
116	Morocco	100
117	Fiji	98
118	Nicaragua	93
119	Samoa	92
120	Burkina Faso	91
121	Bhutan	90
122	Egypt	86
123	Myanmar	81
124	Micronesia	81
125	Solomon Islands	80
126	Congo	79
127	Philippines	78
128	Tonga	77
129	Zimbabwe	66
130	Lesotho	59
131	Ghana	59
132	Timor-Leste	56
133	Cuba	56
134	Vanuatu	55
135	Tajikistan	50
136	Pakistan	50
137	Yemen	49
138	Chad	49
139	Kenya	45
140	Nepal	42
141	Marshall Islands	40
142	Zambia	37
143	Guinea-Bissau	37
144	Kiribati	34
145	Nigeria	33
146	Uganda	33
147	Tanzania	31
148	Gambia	29
149	Ivory Coast	29
150	Senegal	28
151	Angola	27
152	Malawi	27
153	Congo	25
154	Afghanistan	21

Rank	Country	Vehicles per 1,000 People
155	Mozambique	21
156	Guyana	20
157	Mali	19
158	Niger	18
159	Bangladesh	13
160	Papua New Guinea	13
161	Sierra Leone	11
162	Eritrea	11
163	Madagascar	10
164	Rwanda	9
165	Saint Lucia	9
166	Togo	9
167	Sudan	8
168	CAF	8
169	Somalia	6
170	Ethiopia	5
171	Benin	3
172	Guinea	3

Index

Benzene
    Uses for, 32
Brent
    Benchmark, 113
    Composition, 83
    Declining production, 154
    Molecular composition, 84
    Pricing trend, 140
    Quality, 25
    WTI price differential, 153
Butadiene
    Uses for, 32
China
    Dalian refinery, 200
    Fujian refinery, 161
    Petrochina, 213
Crude oil, uses of, 26
Delek Refineries
    Big Spring, 178
    El Dorado, 176
    Krotz Springs, 179
    Tyler, 177
Density, 23
Diesel
    Cetane Number, 81
    Crude Oil Yield, 72
    Definition, 76
    Demand percent of total, 77
    ExxonMobil Sales of, 162
    Market Crack Spread, 111
    Petrochina Sales of, 216
    Saudi Aramco Sales of, 207
    Total SA Production of, 199
    Worldwide demand, 77
Ethylene
    Current costs of production, 94
    ExxonMobil, 163

ExxonMobil Beaumont Plant, 159
ExxonMobil cost advantage, 163
    Petrochina, 216
    Sources of ethylene, 94
    Uses for, 32
ExxonMobil Refineries
    Altona, 161
    Antwerp, 160
    Augusta, 161
    Baton Rouge, 159
    Baytown, 158
    Beaumont, 159
    Billings, 159
    Fawley, 161
    Fos-sur-Mer, 160
    Fujian, 161
    Gravenchon, 161
    Joilet, 160
    Jurong/PAC, 161
    Karlsruhe, 161
    Nanticoke, 160
    Rotterdam, 161
    Sarnia, 160
    Slagan, 161
    Sriracha, 161
    Strathcona, 160
    Trecate, 161
    Yanbu, 162
Gasoline
    Additives, 125
    Components, 75
    Definition, 75
    Demand, 76
    ExxonMobil Sales of, 162
    Fuel Taxes, 133
    Octane Rating, 80
    Saudi Aramco Sales of, 207

Gross Margin, Midstream, 55
Heavy Oil
    Crude Oil Names, 25
    Density, 23
    Jazan Refinery, 209
    Refining, 106
Hydrocarbons
    Types of, 74
Marketing
    Branded, 125
    Crude Oil, 26
    Overview, 31
    Retail, 123
    Unbranded, 127
    Wholesale, 123
Natural gas, 27
    Cubic meter, 39
    Impacts to Refining, 93
    Marketing, 146
NGL, 28
NGL Fractionation
    Saudi Aramco – Ju'aymah, 212
    Saudi Aramco - Yanbu', 212
Norway
    ExxonMobil refinery, 161
Offshore Field
    Karan, 212
Oil, 21
    Composition, 22
    Density, 22
    Futures Contract, 140
    Marketing, 26
    Measuring units, 38
    Metric ton, 39
    Sample crude oils, 25
    Value Chain, 25
Paraxylene
    ExxonMobil, 163
Petrobras
    Brazil Refined Product Sales, 227
Petrobras Refineries
    Abreu e Lima, 225

Alberto Pasqualini, 224
Capuava, 225
Duque de Caxias, 224
Gabriel Passos, 225
Henrique Lage, 224
Isaac Sabbá, 225
Landulpho Alves, 224
Lubricantes e Derivados do Nordeste, 226
Pasadena Refining System, 225
Paulínia, 223
Potiguar Clara, 226
Presidente Bernardes, 224
Presidente Getúlio Vargas, 224
Petrochemicals
    Crude oil, 27
    Downstream overview, 31
    ExxonMobil, 163
    ExxonMobil Worldwide Capacities, 163
    IOCs, 36
    NGL, uses for, 29
    Overview, 31
    Rabigh complex, 212
    Rosneft, 193
    Saudi Aramco Joint Ventures, 210
    TOTAL S.A., 201
Petrochina Refineries
    Dalian, 214
    Dushanzi, 215
    Fushun, 214
    Guangxi, 215
    Jilin, 215
    Lanzhou, 214
    Sichuan, 215
    Yunnan, 215
Polyethylene
    ExxonMobil, 163
Polypropylene
    ExxonMobil, 163
Propylene
    Uses for, 32

Qatar
   Qatar Petroleum, 35
   Total Ras Laffan, 200
Refinery
   ExxonMobil, 158
   Saudi Aramco, 209
   TOTAL SA, 198
Rosneft Refineries
   Achinsk, 191
   Angarsk, 190
   Bashneft, 189
   Bayernoil, 190
   Komsomolsk, 191
   Kuibyshev, 192
   Mini-Refineries, 192
   MiRO, 190
   Novokuibyshev, 191
   PCK, 190
   Ryazan, 189
   Saratov, 191
   Syzran, 191
   Tuapse, 190
Russia
   Rosneft, 187
Saudi Arabia
   Saudi Aramco, 207
Service Stations
   ExxonMobil, 162
   Number of, 124
   Petrobras, 228
   Petrochina, 217
   Rosneft, 192

Tesoro, 171
Total SA, 201, 202
Supply & Trading
   Crude Oil, 139
   Natural Gas Liquids, 144
   Refined Products, 142
Tesoro Refineries
   Anacortes, 168
   Dickinson, 169
   El Paso, 169
   Gallup, 169
   Kenai, 169
   Los Angeles, 170
   Mandan, 170
   Martinez, 170
   Salt Lake City, 170
   St. Paul Park, 170
Toluene
   Use in World War II, 91
   Uses for, 32
Total Refineries
   African, 200
   Asia & Middle East, 200
   France, 199
   Other European, 199
   United States & Other, 200
West Texas Intermediate (WTI)
   Brent price differential, 153
   NYMEX, 140
Xylene
   Uses for, 32

Glossary

AGO: Atmospheric Gas Oil is the heaviest product boiled by a crude distillation unit operating at atmospheric pressure. This fraction ordinarily sells as distillate fuel oil, either in pure form or blended with cracked stocks. In blends atmospheric gasoil, often abbreviated AGO, usually serves as the premium quality component used to lift lesser streams to the standards of saleable furnace oil or diesel engine fuel. Certain ethylene plants, called heavy oil crackers, can take AGO as feedstock

Alkylation Unit: A refinery unit utilizing an acid catalyst to combine smaller hydrocarbon molecules to form larger molecules in the gasoline boiling range to produce a high octane gasoline blendstock, which is referred to as alkylate.

Aromatic: an organic compound characterized by multiple double bonds and a ring structure, like the benzene ring. Aromatics are very stable because the double bonds are conjugated, meaning that the carbon-to-carbon bonds go in a pattern of single bond, then double bond, then single, then double, etc.

Associated gas: Associated gas is natural gas found in contact with or dissolved in crude oil in the reservoir. It can be further categorized as Gas-Cap Gas or Solution Gas.

ATB: Atmospheric Tower Bottoms, also known was Heavy Fuel Oil (HFO) is the undistilled fraction in an atmospheric pressure distillation of crude oil. Also called residual fuel

Barrel/BBL: Volume unit corresponding to 42 U.S. gallons or 159 liters. A U.S. barrel is widely used in the industry

Benzene: a colorless, volatile, flammable liquid used extensively in organic chemistry as a base structure to which different atoms and molecular structures can be attached. It is used to make medicine, crop protection chemicals and many other beneficial products. It is also used as a solvent and component in motor fuels.

Butane-Butylene Fraction: Byproduct of refining processes, typically of cracking different fractions in the FCC unit which produce olefins and light gases including butane-butylene fractions

Biodiesel: A renewable fuel produced from vegetable oils or animal fats that can be blended with petroleum-derived diesel to produce biodiesel blends for use in diesel engines. Pure biodiesel is referred to as B100, whereas blends of biodiesel are referenced by how much biodiesel is in the blend (e.g., a B5 blend contains five volume percent biodiesel and 95 volume percent ULSD).

Blendstocks: Various products or intermediate streams that are combined with other components of similar type and distillation range to produce finished gasoline, diesel fuel or other refined products. Blendstocks may include natural gasoline, hydrotreated Fluid Catalytic Cracking Unit gasoline, alkylate, ethanol, reformate, butane, diesel, biodiesel, kerosene, light cycle oil or slurry, among others.

BOE Barrel of Oil Equivalent: BOE is used as a standard unit to measure combined oil and natural gas. The latter is converted from standard cubic meters into barrels of oil equivalent using a certain coefficient

BPD: Barrels Per Day

Brent Crude Oil: a light, sweet crude oil, though not as light as WTI. Brent is the leading global price benchmark for Atlantic basin crude oils.

BTU: British Thermal Unit. A BTU is equivalent to the energy required to heat 1 pound of water by one degree Fahrenheit. A BTU can also be defined as the amount of energy released by striking a single wooden match.

BTX: Benzene, Toluene, Xylene. Produced as a byproduct of certain refining processes, in particular the reformer unit. A BTX unit in a refinery typically separates these mixture of products into purity products and are either processed by a chemical plant inside the refinery or sold to chemical facilities nearby that can transform these feedstocks into various petrochemicals.

CARB: California Air Resource Board

CBOB: Motor gasoline blending components intended for blending with oxygenates, such as ethanol, to produce finished conventional motor gasoline.

CCR: Continuous Catalytic Regenerator, unit associated with the Reformer where catalyst is transferred from one stage to the other to be re-generated so that it can be used in the catalytic reforming process again.

Complexity Index: A measure of secondary conversion capacity of a refinery relative to its primary distillation capacity. Generally, more complex refineries have a higher index number.

Condensates: Condensate is a mixture of hydrocarbons that exists in the gaseous phase at original reservoir temperature and pressure, but when produced, is in the liquid phase at surface pressure and temperature.

Conversion capacity: Maximum amount of feedstock that can be processed in certain dedicated facilities of a refinery to obtain finished products. Conversion facilities include catalytic crackers, hydrocrackers, visbreaking units, and coking units.

Conversion index: Ratio of capacity of conversion facilities to primary distillation capacity. The higher the ratio, the higher is the capacity of a refinery to obtain high value products from the heavy residue of primary distillation.

Cracking: The process of breaking down larger hydrocarbon molecules into smaller molecules using catalysts and/or elevated temperatures and pressures.

Crude Distillation Capacity, Nameplate Capacity or Production Capacity: The maximum sustainable capacity for a refinery or process unit for a given feedstock quality and severity level, measured in barrels per day.

Crude Oil: Oil is a mixture of molecules of hydrogen and carbon that are primarily found in liquid state at atmospheric conditions. The hydrocarbon mixtures found in crude oil range in properties, such as boiling points and number of carbon and hydrogen molecules.

Cubic Foot: a unit of measure for volume, usually used to measure natural gas. It is the space or volume a cube of 1 by 1 by 1 occupies or in other words 1 foot long by 1 foot wide by 1 foot in height (LxWxH). Since a cubic foot is a very small amount of gas volume, gas is more commonly measured in terms of one thousand cubic feet or 1MCF

Cubic Meter: An international metric system unit of measure for volume, commonly used for measuring natural gas. One cubic meter is equivalent to 35.31 cubic feet

DAO: De-Asphalted Oil

DCS: Distributed Control Systems, system in the refinery or other plant that controls processes in a facility.

Delayed Coking Unit (Coker): A refinery unit that processes or "cracks" heavy oils, such as the bottom cuts of crude oil from the crude or vacuum units, to produce blendstocks for light transportation fuels or feedstocks for other units and petroleum coke.

DHT: Diesel Hydrotreater

Downstream: Usually refers to refining and marketing of hydrocarbons as well as the production and marketing of petrochemicals.

DSU: Desulfurization Unit

E-10: A 90% gasoline-10% ethanol blend.

E-15: An 85% gasoline-15% ethanol blend.

E-85: A blend of gasoline and 70%-85% ethanol.

EIA: United States Energy Information Agency

EPA: Environmental Protection Agency. EPA regulations impact many areas in the oil & gas industry, particularly in setting renewable standard requirements, gasoline composition and other areas.

ESP: Electrostatic Precipitator

Ethanol: An oxygenated blendstock that is blended with sub-grade (CBOB) or conventional gasoline to produce a finished gasoline.

Ethylene: a colorless, flammable, gas that contains only two carbons that are doubly-bonded to one another. It is one of the most important olefins and used extensively in chemical synthesis and to make many different kinds of plastics, such as the plastic used for water bottles.

Exchange Agreement: An agreement providing for the delivery of crude oil or refined products to/from a third party, in exchange for the delivery of crude oil or refined products to/from the third party.

Exploration of oil and natural gas: Exploration that includes land surveys, geological and geophysical studies, seismic data gathering and analysis and well drilling.

Feedstocks: Crude oil and petroleum products used as inputs in refining and petrochemical processes.

FERC: The Federal Energy Regulatory Commission.

Fischer-Tropsch process: One of the most widely used Gas-to-Liquids processes, which converts synthetic gas into valuable liquid petroleum products such as ultra-low sulfur diesel and other fuels.

Fluid Catalytic Cracking Unit: or FCC unit is a refinery unit that uses fluidized catalyst at high temperatures to crack large hydrocarbon molecules into smaller, higher-valued molecules (LPG, gasoline, LCO, etc.).

Fractionation: The process of separating natural gas liquids into its component parts by heating the natural gas liquid stream and boiling off the various fractions in sequence from the lighter to the heavier hydrocarbon.

GAAP: Generally Accepted Accounting Principles. Usually meant to refer to either United States Generally Accepted Accounting Principles (U.S. GAAP) or International Financial Reporting Standards (IFRS). Companies can usually disclose both GAAP and non-GAAP measures depending on the audience and document that is used for.

GDU: Gasoline Desulfurization Unit

GHT: Gasoline Hydrotreater

GOHT: Gas Oil Hydrotreater

GTL: Gas to Liquids. See Fischer-Tropsch process. Gas to liquids technologies convert natural gas into valuable refined petroleum products such as diesel or jet fuel

Gulf Coast 5-3-2 crack spread: A crack spread reflecting the approximate gross margin resulting from processing one barrel of crude oil into 3/5 of a barrel of gasoline and 2/5 of a barrel of high sulfur diesel, utilizing the market prices of WTI crude oil, U.S. Gulf Coast Pipeline CBOB and U.S. Gulf Coast Pipeline

H2: Hydrogen

H2S: Hydrogen Sulfide

HAGO: Heavy Atmospheric Gas Oil

HCU: Hydrocracker Unit

HDS: Hydrodesulfurization

HDT: Hydrotreating

HF: Hydroflouric (acid)

HGO: Heavy Gas Oil

HOC: Heavy Oil Cracker (FCC)

HVGO: Heavy Vacuum Gas Oil

Hydrocracking: A process that uses a catalyst to crack heavy hydrocarbon molecules in the presence of hydrogen. Major products from hydrocracking are distillates, naphtha, propane and gasoline components such as butane.

Hydrotreating Unit: A refinery unit that removes sulfur and other contaminants from hydrocarbons at high temperatures and moderate to high pressure in the presence of catalysts and hydrogen. When used to process fuels, this unit reduces the sulfur dioxide emissions from these fuels.

Inorganic: not containing the carbons and hydrogens bound together like those found in organic compounds. Class of chemicals that typically exist as salts, acids and alkalines, as well as certain gases and elemental compounds. Carbon dioxide, although it contains a carbon atom, is considered inorganic because of its lack of hydrogens bound to the carbon atom.

Isomerization Unit: A refinery unit altering the arrangement of a molecule in the presence of a catalyst and hydrogen to produce a more valuable molecule, typically used to increase the octane of gasoline blendstocks.

Jobbers: Retail stations owned by third parties that sell products purchased *from* a or *through* a downstream company.

LCO: Light Cycle Oil

LGO: Light Gas Oil

Light/Medium/Heavy Crude Oil: Terms used to describe the relative densities of crude oil, normally represented by their API gravities. Light crude oils (those having relatively high API gravities) may be refined into a greater amount of valuable products and are typically more expensive than a heavier crude oil.

LNG: Liquefied Natural Gas is obtained through the cooling of natural gas to minus 260 degrees F at normal pressure. The gas is liquefied to allow transportation from the place of extraction to the sites at which it is transformed back into its natural gaseous state (re-gasified) and then used by consumers for different purposes.

LPG: Liquefied Petroleum Gas usually meant to describe a combination of propane and butanes. LPG is a key source of cooking fuel around the world where there is no natural gas (methane gas) pipeline distribution.

LSD: Low Sulfur Diesel

LSR: Light straight run naphtha.

Margin: The difference between the average selling price and direct acquisition cost of a finished product or raw material excluding other production costs (e.g. refining margin, margin on distribution of natural gas and petroleum products or margin of petrochemical products). Margin trends reflect the trading environment and are, to a certain extent, a gauge of industry profitability.

MCF: One thousand cubic feet. M stands for Roman numeral M or mille for one thousand.

Metric Ton: A measure of mass or weight. A metric ton in this book is converted using a 7.33 barrel per metric ton conversion. The conversion from metric ton to barrel has to factor in the density of the referenced liquid.

Midstream: Usually refers to intermediate processing and transportation of oil & gas products, whether in raw state, intermediate or finished state.

MMBOED: millions of barrels of oil-equivalent per day. Natural gas is converted to a barrel-oil equivalent using a 6,000 cubic feet per barrel conversion factor.

MMBTU: Million BTU. One M stands for Roman numeral M; therefore two Ms stand for one million.

MON: Motor Octane Number

MSCF/d: Abbreviation for a thousand standard cubic feet per day, a common measure for volume of gas.

MTBE: Methyl Tertiary:Butyl Ether

Naphtha: A hydrocarbon fraction that is used as a gasoline blending component, a feedstock for reforming and as a petrochemical feedstock.

Naphthenic: any of various volatile, often flammable, liquid hydrocarbon mixtures characterized by saturated ring structures that are used chiefly as solvents and diluents.

Natural gas liquids (NGL): Liquid or liquefied hydrocarbons recovered from natural gas through separation equipment or natural gas treatment plants. Ethane, propane, normal-butane and isobutane, isopentane and pentane plus are natural gas liquids. NGLs can also be refined from crude oil at a refinery.

Natural Gas: a mixture of hydrocarbons, primarily found in gaseous form. Natural gas in raw form is primarily composed of methane gas, but can also contain ethane, propane, iso-butane, normal butane and pentanes and heavier molecules. When most people refer to natural gas, they are usually referring to "pipeline quality" gas, which is primarily methane and a little bit of ethane, since most of the heavier molecules would have to be removed to prevent "slugs" of liquids clogging and compromising a pipeline's safe operations.

New York Mercantile Exchange (NYMEX): A major commodities futures exchange, where contracts for commodities such as crude oil, natural gas and refined products are bought and sold every day.

Non-GAAP: Any financial measure, used by management that is not part of either U.S. GAAP or IFRS. These non-GAAP financial measures, such as gross margin, are commonly used internally by companies to track performance. These non-GAAP measures are usually derived from adjustments from GAAP measures. For more information, please visit: http://www.sec.gov/rules/final/33-8176.htm

NOX: Nitrogen Oxides

Olefin (aka Alkene): hydrocarbons characterized by having at least one double bond; specifically, any of a series of open-chain hydrocarbons such as ethylene.

Organic: chemical compounds containing carbon atoms bonded to other carbon atoms, hydrogen atoms or other substitutes for hydrogen (e.g., halogens, sulfur, nitrogen, etc.)

Propylene: a three-carbon, flammable, gaseous molecule containing a double-bond; another important olefin used in organic synthesis. Propylene is also a base chemical to make polypropylene fibers, which are used in high-performance clothing, carpeting and other products.

Operating margin: net sales less cost of goods sold.

P:P: Propane:Propylene

Per barrel of sales: calculated by dividing the applicable income statement line item (operating margin or operating expenses) by the total barrels sold during the period.

Petroleum Administration for Defense District (PADD): Any of five regions in the United States as set forth by the Department of Energy and used throughout the oil industry for

Petroleum Coke: A coal-like substance produced as a byproduct during the Delayed Coking refining process.

Primary balanced refining capacity: Maximum amount of feedstock that can be processed in a refinery to obtain finished products measured in BBL/d.

PSI: Pounds per Square Inch

RBOB: Reformulated Blendstock for Oxygenate Blending

Refining margin, refined product margin or crack spread: A metric used in the refining industry to assess a refinery's product margins by comparing the difference between the price of refined products produced at the refinery and the price of crude oil required to produce those products.

Reforming Unit: A refinery unit that uses high temperature, moderate pressure and catalyst to create petrochemical feedstocks, high octane gasoline blendstocks and hydrogen.

Renewable Fuels Standard 2 (RFS-2): An EPA regulation promulgated pursuant to the EISA, which requires most refineries to blend increasing amounts of renewable fuels (including biodiesel and ethanol) with refined products.

Renewable Identification Number (RIN): a renewable fuel credit used to satisfy requirements for blending renewable fuels under RFS-2.

RFG: Reformulated Gasoline

RON: Research Octane Number

Roofing flux: An asphalt-like product used to make roofing shingles for the housing industry.

RVP: Reid Vapor Pressure

SEC: United States Securities & Exchange Commission is an independent agency of the U.S. Federal Government that oversees securities exchanges and enforces U.S. federal securities laws. The SEC regulates company issued documents, such as form 10-Ks, 20-Fs, annual reports, earnings releases and other externally published documents.

Ship-or-pay: Clause included in natural gas transportation contracts according to which the customer is requested to pay for the transportation of gas whether or not the gas is actually transported.

SOX : Sulfur Oxides

SRU: Sulfur Recovery Unit

Straight run: product produced off of the crude or vacuum unit and not further processed.

Synthesis gas: a mixture of carbon monoxide and hydrogen used especially in chemical synthesis to make hydrocarbons.

Sweet/Sour crude oil: Terms used to describe the relative sulfur content of crude oil. Sweet crude oil is relatively low in sulfur content; sour crude oil is relatively high in sulfur content. Sweet crude oil requires less processing to remove sulfur and is typically more expensive than sour crude oil.

Take-or-pay: Clause included in natural gas supply contracts according to which the purchaser is bound to pay the contractual price or a fraction of such price for a minimum quantity of gas set in the contract whether or not the gas is collected by the purchaser. The purchaser has the option of collecting the gas paid for and not delivered at a price equal to the residual fraction of the price set in the contract in subsequent contract years.

TAN: Total Acid Number

Throughput: The quantity of crude oil and feedstocks processed through a refinery or a refinery unit.

Toluene: a liquid aromatic hydrocarbon that has a benzene-like structure but is less volatile, flammable, and toxic than benzene.   Toluene is used in organic synthesis, as a solvent, and as an antiknock agent for gasoline.

Turnaround: A periodic shutdown of refinery process units to perform routine maintenance to restore the operation of the equipment to its former level of performance. Turnaround activities normally include cleaning, inspection, refurbishment, and repair and replacement of equipment and piping. It is also common to use turnaround periods to change catalysts or to implement capital project improvements.

U.S. Gulf Coast Pipeline CBOB: A grade of gasoline blendstock that must be blended with 10% biofuels in order to be marketed as Regular Unleaded at retail locations.

U.S. Gulf Coast Pipeline No. 2 Heating Oil: A petroleum distillate that can be used as either a diesel fuel or a fuel oil. This is the standard by which other Gulf Coast distillate products (such as ultra-low sulfur diesel) are priced.

Ultra-Low Sulfur Diesel (ULSD): Diesel fuel produced with a lower sulfur content (15 ppm) to reduce sulfur dioxide emissions. ULSD is the only diesel fuel that may be used for on-road and most other applications in the U.S.

Upstream/Downstream: The term upstream refers to all hydrocarbon exploration and production activities. The term downstream includes all activities inherent to the oil and gas sector that are downstream of exploration and production activities.

Vacuum Distillation Unit: A refinery unit that distills heavy crude oils under deep vacuum to allow their separation without coking.

VGO: Vacuum Gas Oil, byproduct of crude oil vacuum distillation that is typically sent to a hydro-cracking unit for upgrading.

VOC: Volatile Organic Compound, are organic compounds that easily become vapors or gases and are released from burning fuel, but also released from solvents, paints, glues and other products.

VTB: Vacuum Tower Bottoms, are the left over bottom product of distillation, which can be processed in cokers and used for upgrading into gasoline, diesel and gas oil.

West Texas Intermediate Crude Oil (WTI): A light, sweet crude oil characterized by an API gravity between 38 and 44 and a sulfur content of less than 0.4 weight percent that is used as a benchmark for other crude oils.

Xylenes: one of the major aromatic feedstocks that is usually obtained from petroleum or natural gas distillates. Xylenes are used in the manufacture of plastics and synthetic fibers, as a solvent and in the blending of gasoline.